Dealing with Genetically Modified Crops

Editors

Richard F. Wilson
USDA ARS
Raleigh, North Carolina

Ching T. Hou
USDA ARS NCAUR
Peoria, Illinois

David F. Hildebrand
University of Kentucky
Lexington, Kentucky

AOCS PRESS

AOCS Mission Statement
To be a forum for the exchange of ideas, information, and experience among those with a professional interest in the science and technology of fats, oils, and related substances in ways that promote personal excellence and provide high standards of quality.

The paper used in this book is acid-free and falls within the guidelines established to ensure permanence and durability.

Library of Congress Cataloging-in-Publication Data

Dealing with genetically modified crops / editors, Richard F. Wilson, Ching T. Hou, David F. Hildebrand.

p. cm.

ISBN 1-893997-22-7

1. Crops--Genetic engineering--Congresses. I. Wilson, Richard F., 1947- II. Hou, Ching T. (Ching-Tsang), 1935- III. Hildebrand, David F. IV. American Oil Chemists' Society. Meeting (91st : 2000 : San Diego, Calif.)

SB123.57 .D43 2001
631.5′233--dc21 2001022337
CIP

Printed in the United States of America with vegetable oil-based inks.
00 99 98 97 5 4 3 2 1

Preface

A special symposium entitled, *Dealing with Genetically Modified Crop* was presented at the 91st annual meeting of the American Oil Chemists' Society held in San Diego, CA on April 25–28, 2000. This symposium was convened during a period of time in which the U.S. economy was growing steadily at a rapid rate, but this apparent prosperity was not shared with segments of the oilseed processing industry. Indeed, many companies experienced extremely tight "margins," and many U.S. farmers produced crops below cost. As an example, the U.S. farm price for soybeans had fallen to ~$4.36/bu, a 20 year low. Although this dilemma pertained largely to the major commodities, corn and soybeans, it also rippled through the markets for other crops such as peanut, sunflower, and rapeseed. After achieving near record price highs in 1997, the initial estimated price for these crops (Rotterdam or Hamburg CIF) in 2000 had declined, i.e., –$219/MT for peanut, –$129/MT for sunflower, and –$111/MT for rapeseed. Many economists suggested that this market environment was a function of accelerated growth in commodity supply, and the evidence seemed to support this theory. In the case of soybeans, world supply grew ~45 MMT between the years 1996 and 2000, whereas demand grew only ~35 MMT, resulting in about a 10 MMT growth in end-stocks. However, the symptoms of this malady were confined almost exclusively to the United States (Table 1). Excluding the United States from this calculation, growth in soybean supply from all other sources apparently equaled growth in demand. Further inspection revealed that a major contributing factor to the estimated growth in excess supply of U.S. soybeans was essentially the zero growth in U.S. soybean exports. In fact there was a precipitous loss in the U.S. share of global soybean exports during this period, from 72 to ~57% (Fig. 1). In particular, this episode of poor performance in export markets was vexing to U.S. soybean producers. Not only had South American soybean production captured the lost U.S. share in the global export market, but there was also a strong negative correlation ($R^2 = 0.87$) between the farm price for U.S. soybeans and the relative U.S. share of world soybean exports in this 4 year period.

TABLE 1
Growth in the Global Soybean Market

	Change between 1996 and 2000 in MMT		
Origin	Supply[a]	Demand[b]	Excess
USA	16.8	7.3	9.5
All other	27.8	27.7	0.1
Total	44.6	35.0	9.6

[a]Difference in annual production plus end-stocks.
[b]Difference in annual crush plus export.
Source: Foreign Agricultural Service, Oilsseds:World Markets, Circular Series, FOP 11-00, U.S. Department of Agriculture, Washington, DC, 2000.

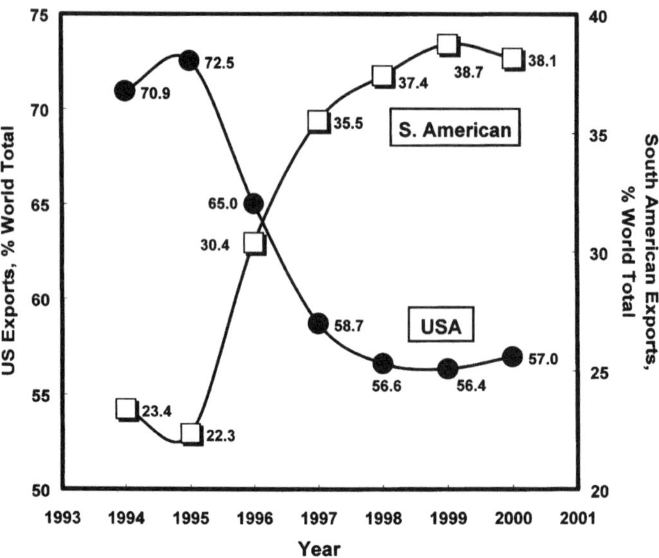

Fig. 1. Trends in world soybean export.

Because soybeans have a major effect on the price structure for other agricultural crops, it may be assumed that a more equitable balance among nations in ability to service demand is paramount to the economic recovery of all commodity markets. This brings us to the reason for this symposium. One cannot fail to notice the coincidence of the downward turn in commodity prices with the emergence of transgenically modified (GMO) crops in world trade. Obviously, the oilseed industry is concerned by the issues surrounding the debate on use of GMO crops in edible products. Uncertainties about their application continue to tear at the fabric of U.S., as well as global oilseed trade. Although it is important to encourage the development of this science, it is inevitable that rapid deployment of any new technology will be accompanied by unintentional mistakes. Associated liabilities from these unfortunate events may subsequently permeate the system and settle the issue. However, the process for regaining consumer confidence in agricultural products should not rely upon or be subjugated by legal remedies. In that regard, U.S. Secretary of Agriculture Glickman has taken first steps to define the U.S. approach to biotechnology in the 21st century. Following his lead, this symposium was designed to promote understanding and expand dialogue on how to deal with GMO crops. By providing a neutral venue, it was hoped that these discussions would help enhance credibility among international agricultural policy makers and industrial associations or companies involved in trade or manufacture of agricultural products. Significant progress toward this goal was enabled by the admirable presentations given by distinguished representatives from the following groups: Department of Health of Taiwan; Department of Trade and Industry of England; Ministry of Agriculture, Fisheries and Forestry of Japan; Federal Research Centre for Nutrition of Germany; Agriculture and Agrifood Canada; and the EC Seed Crushers and Oil Processors Association of Belgium. In particular, these presentations revealed that governments of the represented countries were not entirely against GMO crops. However, there also was consensus that a global debate should be held to give full consideration to the nuances of this complex situation. Such a forum should be, in the words of one speaker, "One that might help to clear up some of the regulatory confusion associated with too many rules and protocols. Also, industry and government must listen to the consumer and respond, in kind, being careful not to alienate buyers."

"Listening to consumers and responding in kind" was the recurrent theme throughout the first session of this symposium. Yet this challenge is a responsibility that must be shared not only by industry and government, but by science as well. In this regard, it is important for scientists to help society understand what biotechnology encompasses. Too often the term "biotechnology" is associated exclusively with the study and application of transgenic events. This misconception overlooks a majority of the tangible beneficial contributions to this science that come from the areas of genomics, proteomics and Mendelian genetics, all of which are essential components of biotechnology. Therefore, another goal of this symposium was to present a balanced view of genetic technologies that are required for future advances in agriculture. To help gain a better appreciation of plant genetics, the second session of this symposium featured technical discussions on how crop improvement may be accelerated by exploration of plant genomes, use of natural genetic diversity, characterization of gene function, mechanistic knowledge of gene interactions, development and use of specific gene markers in conventional breeding, and new methods to exploit or create variation in natural genes for enhanced crop quality or productivity. These topics were addressed fully by the Administrator of the USDA-Agricultural Research Service and eminent scientists engaged in soybean, corn, rice, rape, or cotton research. In addition, this session showcased the most current commercial processes used to detect transgenic DNA in food products, previewed what may yet come in the way of trans-gene technologies, and accentuated even more recent biotechnological advances that provide the ability to develop varieties with natural genetic modifications for improved crop value and productivity.

In the final outcome, ways must be found to help keep market options open and enhance the competitive position of agricultural crops in world trade. Progress toward this goal will require careful consideration of the purpose and means for making genetic changes in any food crop, and continued discussion among all parties involved. As reflected in this *Proceedings*, the symposium *Dealing with Genetically Modified Crops* took a small but positive step forward in this process. For these reasons, the organizers of this symposium gratefully acknowledge the willing contributions of all the speakers and authors. Special appreciation also is noted for the financial assistance provided by the USDA Agricultural Research Service and the American Oil Chemists' Society, which helped bring these individuals together.

Richard F. Wilson
Ching T. Hou
David F. Hildebrand

Contents

Chapter 1

Regulation of Genetically Modified Organisms in Taiwan

Shu-Kong Chen, Lu-Hung Chen, and Hung-Yi Su

Bureau of Food Sanitation, Department of Health, The Executive Yuan, Taiwan, R.O.C.

Introduction

Biotechnology has been widely applied by industries recently to produce various products including foods, medicine, and other agricultural produce in Taiwan. An estimate of $0.5 billion of production value in the bioindustry was reported in 1997; this value is expected to reach $2.5 billion in 2005. However, this estimation of progress in biotechnology excludes genetically modified (GM) foods. Therefore, with the addition of GM foods, a higher production value can be expected in the future.

Traditional selective breeding approaches to improve a production trait are limited because the progress in genetic gain is slow. Because genetic information is involved in whole chromosome exchange, one desirable production trait is difficult to separate from others (1,2). In contrast, genetic modification using modern biotechnology enables gene transfer to introduce a single gene of exactly defined sequence and to allow genetic information to be transferred between species (3,4). Using this kind of gene transfer, scientists have been able to produce GM foods with the desired traits that are made from plants, animals, or microorganisms. These GM foods can be whole foods containing new genetic material (e.g., a GM soybean) or processed foods containing ingredients with new genetic material (e.g., tofu produced from GM soybeans) or highly refined foods derived from GM organisms but containing little or no new genetic material (e.g., soybean oil).

Traits of GM Foods

Although production traits of GM foods may be improved more efficiently compared with conventional foods, other traits such as toxicity, allergenicity, or altered composition and nutritional value, which are due to the introduction of a new gene, may emerge. To meet the basic requirement of environmental and food safety, food and food ingredients produced using gene technology that are offered for sale in Taiwan must comply with regulations of different governmental agencies. Among these agencies, the upstream regulator, the National Science Council (NSC), plays a supervisory role in research and development of genetically modified organisms (GMO). The midstream regulators, Environmental Protection

Administration (EPA) and Council of Agriculture (COA), are responsible for environmental protection and field testing, respectively. The downstream regulator, the Department of Health (DOH), is concerned with the safety and labeling of GM foods.

Guidelines and Regulations

Guidelines

In 1988, the "Guideline for Recombinant DNA Experiments" was promulgated by NSC. A key provision of the guideline is that safety must be ensured by conducting these experiments using physical and biological methods of protection. Physical protection restricts recombinants to a defined area or facility, preventing contamination of workers and other facilities. The level of biological protection is set by considering the selection of the genes introduced, the sources and characteristics of the vector, and the contact of recombinants and their products with human beings. The "Guideline for Recombinant DNA Experiments" was designed originally for those receiving grants from NSC, but is now under revision to broaden its application to all research work in the nation. On the basis of the revised proposal under discussion, it is not only scientists undertaking GMO experiments who bear the obligation of following good laboratory practices; NSC also must report any aberration to the corresponding agency if the hazard results from noncompliance.

Regulations

To date, although a majority of academic and industrial scientists agree that some GMO such as transgenic food crops are safe for the environment, others are concerned that certain applications of plant biotechnology are hazardous. In particular, deliberate releases of genetically modified plants are regarded as risky (3). To determine the possible adverse environmental influences due to the emergence of GMO in the environment, the EPA relies on the Environmental Impact Assessment Act that was promulgated in 1994 and amended in 1999. However, this Act has been enforced only when development activity occurs (e.g., to change forest into farmland). The "Environmental Application Agent Management Act," promulgated in 1997, is another law laid down by EPA to protect the environment from contamination with different agents, including environmental sanitation agents as well as water and waste treatment agents. The EPA relies on the "Regulation on Research and Development of Microbial Preparations by Genetic Engineering for Environmental Application," which was especially stipulated in the Environmental Application Agent Management Act to regulate the microorganisms made from gene technology for environmental application. Examples in this regard are those GMO used in water or waste treatment. However, the EPA in Taiwan is not involved in the regulation of chemicals used for pest control in agriculture, a responsibility that belongs to the COA.

To assess the risk that may affect agricultural environments due to the application of GMO, COA relies on the "Guideline for Field Tests of Transgenic Animals" and "Guideline for Field Tests of Transgenic Plants," which were promulgated in 1998. According to both guidelines, any request for field tests regarding transgenic plants and animals originating either domestically or through importation must be made through COA. No deliberate release of GMO in the field and subsequent production for commercial purposes are allowed unless otherwise evaluated and approved by COA. To enforce the risk assessment guidelines more effectively, their legal position may be elevated by incorporating them into the pertinent existing laws. In response to the increasing number of studies on transgenic aquatics in Taiwan, COA is currently discussing whether a draft of the "Guideline for Field Tests of Transgenic Aquatics" is required.

To protect consumers' health and for their interest, DOH is responsible for the regulation of GM food,, based on the "Law Governing Food Sanitation." Because food safety and labeling are two important aspects of GM food regulation, DOH has proposed the "Guidelines for Safety Assessment of Foods from Recombinant DNA Technologies" and is drafting the "Labeling Provisions on GM Foods," respectively. According to the proposed guideline, DOH will evaluate the safety of GM foods by examining both the manufacturing process and the product. This evaluation includes assessment of the safety of the manufacturing methods, materials, equipment and products. In addition, DOH is in the process of identifying proper approaches for GMO labeling because consumers have indicated their desire to be informed whether the foods they consume are or contain GMO. In a market survey of GMO acceptability conducted in 1999, the majority of people interviewed (i.e., ~94% of 1070 people) agreed that labeling is required even under a premarketing approval regulatory regimen (5). This result suggests that specific labeling is an essential part of GMO regulations, and a proper label may be an essential factor in public acceptance of GM foods.

Recent agreement on the "Cartagena Protocol on Biosafety" has also brought attention to the GMO regulators in Taiwan. For example, EPA, COA and DOH will coordinate and draft regulations consistent with the requirement of the Cartagena Protocol to manage the import and export of living modified organisms (LMO). At this moment, it is somewhat unclear how the implementation of this Biosafety Protocol will coexist with rights and obligations already defined under agreements of the World Trade Organization. In addition, the issue of the labeling of LMO is not well addressed by the Biosafety Protocol, thus creating an area of potential controversy. However, the fundamental task of all of the relevant governmental agencies in Taiwan is to lay the groundwork for the effective implementation of the regulation of GMO.

References

1. Ward, K.A., and C.D. Nancarrow, The Genetic Engineering of Production Traits in Domestic Animals, *Experienta 47:* 913–922 (1991).

2. Mazur B., E. Krebbers, and S. Tingey, Gene Discovery and Product Development for Grain Quality Traits, *Science 285:* 372–375 (1999).
3. Ruibal-Mendieta N.L., and F.A. Lints, Novel and Transgenic Food Crops: Overview of Scientific Versus Public Perception, *Transgenic Res. 7:* 379–386 (1998).
4. Hammond J., Overview: The Many Uses and Applications of Transgenic Plants, *Curr. Top. Microbiol. Immunol. 240:* 1–19 (1999).
5. Liang, C.-M., H.-M. Yang, H.-C. Hsieh, K.-W. Hu, and H.-F. Yang, A Market Survey for the Acceptability of Agricultural Biotechnological Products in Taiwan. *Bioindustry 10:* 236–239 (1999). (In Chinese)

Chapter 2

Perspectives on GMO Crops in the UK

Monica Darnbrough

Department of Trade and Industry, London, SW1W 9SS United Kingdom

Introduction

In this paper, my goal is to offer insight into the current status of biotechnology and all aspects of genetic modification, particularly its applications to crops and food, in the UK. However, it is impossible to look at what is happening in the UK without looking also at what is being thought and done in the rest of Europe. It is important to understand the context for this discussion.

In December 1999, the UK Government published a report entitled Genome Valley; it was a report endorsed by all Ministers with an interest in biotechnology. The report gave a snapshot of the UK position. It looked at bioscience and at how the research was being exploited and used in a wide range of sectors, including agriculture, food, food processing, pharmaceuticals, diagnostics, and other health care applications. The report concluded that the UK was second to the United States in terms of the numbers of small (and growing) biotechnology companies, venture capital investment in the sector, jobs created, and on many other measures. All this is rooted in the very strong science base in the UK. The report made it clear that the Government wants the UK to remain #1 in Europe in biotechnology. However, notwithstanding the enormous potential offered by biotechnology, the focus of the Ministers' attention since then has tended to be on ensuring that public concerns over food safety and environmental protection have been addressed.

The Regulatory Framework

A review of the regulatory framework for all applications of biotechnology was conducted by the Government last year. There was wide public consultation. This review concluded that the UK committees of experts, which already existed to look at specific issues, were working well. There are ~20 specialist committees looking at a wide range of applications from genetic testing to matters that relate to human embryology and fertilization. These include the committee that considers applications to grow crops; it functions under the European Directive on Deliberate Releases to the Environment (90/220) and is called the Advisory Committee on Releases to the Environment (ACRE). There is also a committee that considers applications to market new foods, including those developed using genetic modification (GM) techniques—the Advisory Committee on Novel Foods and Provisions, ACNFP.

The review also concluded that, in addition to these specialist groups, there was a need to find a way to include a wider range of nonspecialist interests in the debate about biotechnology applications. As a consequence, two new commissions have been set up, one on human genetics issues [Human Genetics Commission (HGC)] and one on agriculture and environment issues (the Agriculture and Environment Biotechnology Commission). The HGC has just started its work. It is being chaired by Baroness Helena Kennedy, a lawyer, and it includes people who are experts, as well as consumers and representatives from nongovernment organizations. Its first meeting was held in public on 10 April 2000. Several hundred people attended, and a wide range of issues were raised. The minutes and papers will be made public. This reflects an increasing move toward greater openness and transparency in the work of all the groups in this area. This is taking place against a background of the UK Government just beginning to debate more generally a Bill on open government.

The third newly established entity is the Food Standards Agency which was set up on 1 April, 2000. It will bring greater independence to work on food safety issues, which have been moved out of the Ministry of Agriculture. It is being headed by Prof. Sir John Krebs who chaired the OECD conference on GM food issues, held recently in Edinburgh as follow up to G8 concerns.

In the UK, there are many parts of Government that have an interest in and a responsibility for biotechnology. Starting with the center, the Minister for the Cabinet Office coordinates biotechnology across all Government Departments, including Industry and Trade, Health, Agriculture, Environment, and others. There are regulatory groups looking at detailed issues on a case-by-case basis against specific regulations. Then there are the two new Commissions that include all stakeholders. In addition, there is also the influence of Parliament, both the elected chamber, the Commons, and the Lords. In the last few months, the specialist committees of both Houses have looked into aspects of biotechnology. The House of Lords Science and Technology Committee has looked at nonfood crops. They recommended that greater attention be paid to the potential offered by emerging developments. The Government is obliged to respond to these reports and it is currently consulting on proposals to follow up the Lords' ideas. One action will be to set up an industry/Government forum, effectively to do horizon-scanning, looking at emerging science and thinking about how it might be used in new products. It will also advise on R&D priorities as well as on UK and EU policy. It will be chaired by an industrialist, a user, so as to be focused on exploitation and market issues.

The Genome Valley report and the public consultation about regulation identified the need for both more debate in the UK about the social, cultural, ethical, and moral issues arising from the use of biotechnology, and from genetic modification in particular. Several years ago, the UK Government considered whether to set up a Government commission on these matters, as is the case in the United States. It chose, rather, to share the financing of an independent group, the Nuffield Council on Bioethics; the other funders include the Medical Research Council and the private sector funding body, the Wellcome Trust. It has established a reputation as a very professional organization, thus ensuring that it obtains participation in its deliberations from all stakeholders. It has prepared reports on GM food and crops and, very recently, on the use of stem cells.

GM Food and Crops

Against that regulatory framework, what then is the current UK position regarding GM food and crops? First, consider crops. Only four crops have been authorized to be grown under the EU Directive on deliberate release to the environment, i.e., one variety of GM tobacco, two varieties of GM corn, one variety of GM rapeseed oil, and two varieties of GM carnations (for cultivation under glass).

No commercial plantings are in progress in the UK. Field trials are just starting of corn, oilseed rape and beet on sites between 10 and 50 ha, beside control crops, in a variety of locations around the UK. A scientific committee has directed the process, identifying the appropriate size of planting, separation distances, types of soil, and climatic conditions. More than 100 farmers put in applications to carry out a trial. Some of the sites were rejected on scientific grounds because soil or other conditions were unsuitable. We expect to have enough sites to enable valid trials to be done. The crops will be sown this year. Care has also been given to ensure that the crops do not go into human food. There will be safeguards about use of the land after the trial crops have been harvested.

There are also just three GM varieties that have be authorized for the use as food: soy, tomato (used in paste), and corn. These were all approved before the EU Novel Foods Directive came into force in 1997. No new GM food has yet been approved for the UK under this regulation.

In 1998, the UK looked well placed to become a good market for GM crops and food. The UK company, Zeneca, had very successfully introduced the GM tomato (modified for slower ripening, thus giving the advantage that fewer tomatoes were damaged because of their softness after harvesting). The tomato was used to make tomato paste and was sold as such in several major UK supermarket chains. It was competitive on both price and flavor. GM soybeans were being used in a wide range of processed foods, along with oils obtained from the beans.

Then, suddenly, a media campaign blew up. One frozen food retailing chain (Iceland) stated that it would differentiate itself by not selling products containing GM ingredients. The newspapers picked up the story and prolonged for ~2 wk, each day making more and more hysterical headlines. "Frankenstein Foods" was the one that has persisted. Other retail supermarkets followed suit and, very quickly, the tomato paste disappeared from the shelves. Handling products containing soy and soy derivatives was not as easy because these ingredients were pervasive. There was pressure to give consumers information so that they could choose. This demand was felt throughout Europe, and on 10 April, 2000, an EU Regulation to amend foodstuff labeling rules come into force. (The presence of GM soy or corn had had to be indicated since September 1998.) The new Regulation covers mass caterers as well. It is based on a 1% threshold for GM content.

One cannot look at the UK without looking at Europe. At that level, issues are considered by expert groups and then by Ministers from all of the EU member countries. The European Parliament now has strengthened powers and it looks at draft Directives and Regulations and often suggests changes. At the European level, there are a number of issues currently under consideration that could have significant effects on all types of innovation. These are likely to be applied first to GM/environment issues. Examples are the use of the precautionary principle and consideration of liability issues. The European Commission prepared a White Paper on environment liability (going far wider than just GM) last year. Consultation is taking place. However, pressure exists to look specifically at the liability of companies and farmers and at how liability would be handled if there were ever to be a harmful effect of a GM crop on the environment. One possibility is that liability may be considered as an amendment to the Directive on Deliberate Release to Environment, or some Member States may consider taking action at the national level.

Conclusion

What then of the future? In my view, there must be more debate and more importantly, better debate. The debate must be underpinned by the provision of scientifically accurate information, in a form that is easy to understand, and by better education in biotechnology. The social and ethical issues must be addressed. But companies and scientists have a duty to explain and demonstrate the potential benefits to consumers and the public. There must be more research concerning environmental effects. However, basic research programs should also continue while the debate is held. In my personal view, only when we have products with demonstrable benefits that consumers can instantly appreciate or that give real benefits to the developing world will we see the start of acceptance of GM crops in the UK.

Chapter 3

Environmental Risk Assessment of GMO and Activities for Building Public Acceptance in Japan

Yutaka Tabei

Research Council of Agriculture, Forestry and Fisheries, Chiyoda, Tokyo, Japan 100-8950

Introduction

Genetic engineering is increasingly becoming an essential aspect of plant breeding, and various transgenic crops have been commercialized in the United States, Canada, Australia and numerous other countries. In 1994, the long shelf life transgenic tomato, "FLAVR SAVR" was launched in the United States market. In 1999, herbicide-tolerant transgenic soybeans comprised ~50 and 90% of the soybean cultivation areas in the United States and Argentina, respectively (1). During the same years, transgenic corn was grown in ~33% of the total cultivation area in United States, and herbicide-tolerant canola in Canada increased to 62% of the total cultivation area (1). The global farmland used for the cultivation of transgenic crops in 2000 is not expected to exceed that in 1999.

Transgenic crops must be subjected to risk assessment before commercialization. The potential risk of genetically modified organisms (GMO) on the natural environment was first pointed out by scientists during the initial stages of the development of transformation technology. Subsequently, at the meeting of scientists from various countries in California, discussions on the regulation of GMO at the experimental stage led to an agreement that they should be developed and utilized under self-imposed controls (2,3).

In 1976, the National Institutes of Health (NIH) enacted the first guidelines concerning the experimental use of GMO (59FR34496-34547, 1994). Subsequently, in the 1980s, the Organization of Economic Cooperation of Development (OECD) discussed the need for risk assessment before industrial application of GMO, and the committee announced their "Recombinant DNA Safety Considerations" to member countries in 1986. In accordance with this recommendation, the United States, Canada and European countries enacted new items under their existing laws, and organized risk assessment systems for the industrial utilization of GMO. The recommendations of OECD included large-scale industrial applications (tank cultures of microorganisms) and also agricultural/environmental applications (open system applications).

Moreover, OECD has continued to discuss and develop new concepts to evaluate environmental risk and ensure food safety. In 1993, OECD published documents entitled "Safety Considerations for Biotechnology: Scale-up of Crop Plants" (4) and "Safety Evaluation of Foods Derived by Modern Biotechnology: Concepts and Principles" (5), in which two important principles, "familiarity" and "substantial equivalence," were described. The concept of familiarity can be applied to wide aspects, from field trials to seed multiplication of genetically modified crops. An environmental risk assessment must be conducted according to experiments on crop breeding, while the environment surrounding the cultivated genetically modified crop plants must also be taken into consideration. The other concept used for food safety evaluation, substantial equivalence, ensures the food safety of transgenic crops by ensuring the safety of the gene products and comparing the contents of nutrients, toxic substances, and allergens between transgenic crops and original parental lines. These two concepts are now used in many countries as the basis of risk assessment for environmental and food safety.

To promote clear and efficient regulations and to enhance trade of products resulting from biotechnology, the "Harmonization of Regulatory Oversight in Biotechnology" was initiated by OECD in 1997. This working group has continued to publish consensus documents related to the biology of plants, traits introduced into plants, and the biology of microbes to clarify the baseline of risk assessment in member countries. Moreover, the internet homepage (http://www.oecd.org/ehs/service.htm) of the working group provides information on the present conditions of risk assessment in member countries, and provides links to administrative organizations of these countries.

Guidelines for Risk Assessment of Transgenic Crops in Japan

Appropriate risk assessments of the environment, food and feed are required/conducted, depending on the purpose of the transgenic crop production. In Japan, four Ministries and one Agency have their own particular set of guidelines. Guidelines of the Science and Technology Agency (6) address risk assessment of GMO in laboratory use. On the other hand, the environmental risk of GMO falls under the guidelines of the Ministry of Agriculture, Forestry and Fisheries (7), and feed safety is evaluated by another set of MAFF guidelines. Although food safety has been evaluated according to the guidelines of the Ministry of Health and Warfare (MHW), MHW recently changed its policy for risk assessment of food by guideline (Fig. 1). The new framework is for food safety to be evaluated as part of the Food Sanitation Law; this will be enforced from April 1, 2001.

Environmental Risk Assessment System for Transgenic Crop Plants

Environmental risk is evaluated on a step-by-step and case-by-case basis. These steps are classified into the following four levels: Closed greenhouse, semiclosed greenhouse, isolated field, and open field, each with its own particular require-

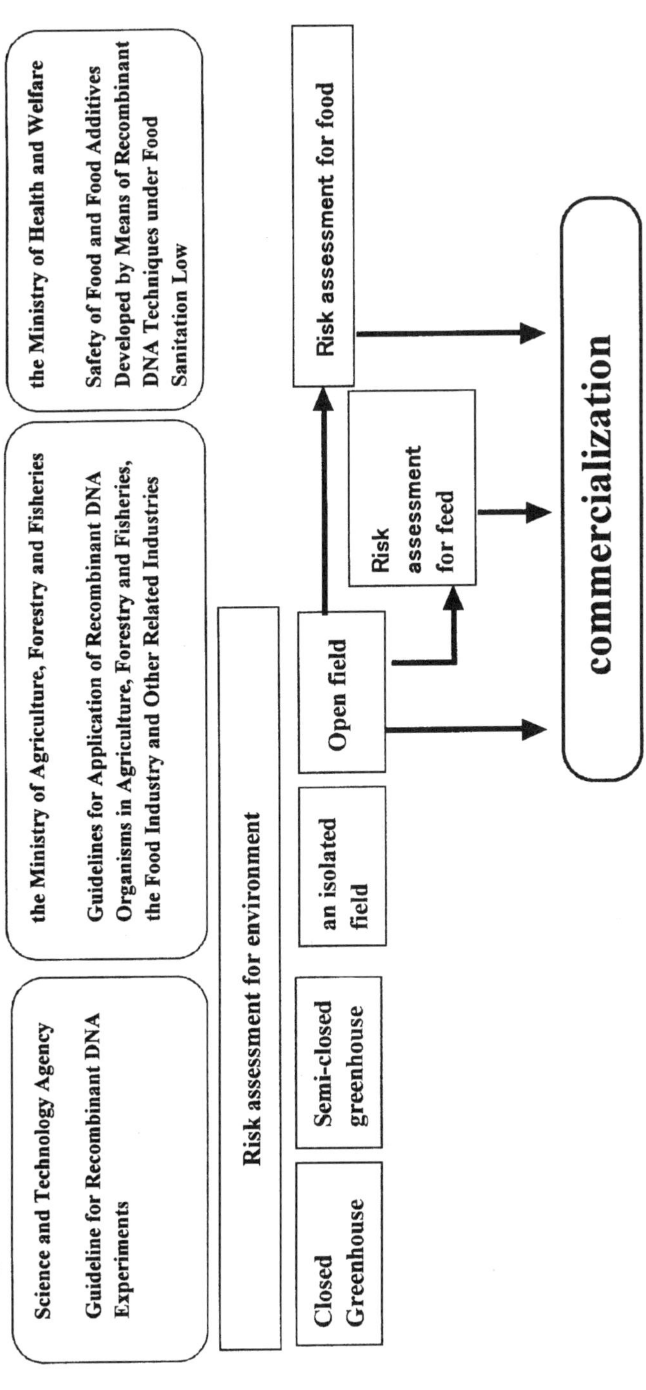

Fig. 1. Framework for the regulation of rDNA crops.

ments. (i) In a closed-greenhouse system, the windows must be completely closed, the air must be ventilated through HAPA filters to avoid scattering pollen and/or dust from genetically modified crop plants to the outside, and the temperature must be controlled by an air conditioner. Moreover, the transgenic plants, the soil and pots used to culture these plants, and the waste drainage may be discarded only after sterilization by autoclaving or by other appropriate methods. (ii) The semi-closed greenhouse system basically requires the same conditions as the closed greenhouse, except that air is allowed to ventilate between the inside of the greenhouse and the outside through open windows which must, however, be covered by mesh (pore size: ~2 mm^2) to avoid invasion and scattering of pollinators. (iii) In an isolated field system, a fence separates the field from ordinary fields. Moreover, an incinerator is required to destroy the transgenic plants after the field trial, and a washing area for the tractors used in the isolated field is necessary. (iv) In the final step of cultivation, genetically modified crop plants can be grown in an open field. At this point, there are no restrictive conditions for the cultivation of transgenic plants, although MAFF requests reports of such tests in the first year of cultivation.

By June 2000, 51 applications including 85 breeding lines were recognized as being biosafe. Twenty-four of 51 can be cultivated in Japan and 24 genetically modified crops can be imported from foreign countries. Moreover, official approval for field trials has been issued to 85 applications including 145 breeding lines (http://ss.s.affrc.go.jp/docs/sentan/index.htm).

Food Safety Assessment

The MHW requests food safety assessment data from the developer. The data should include a comparison of the nutritive value of the transgenic and nontransgenic plants, characteristics of the host plant, the sequence and function of the introduced gene, the traits of the vector, and the characteristics of the new protein encoded by the introduced gene. Moreover, the encoded protein must be evaluated for the possibility that it possesses allergen activity and toxicity, such as by intake studies and comparisons of the sequence to known allergens and toxins. Furthermore, the sensitivity of the protein to heating or artificial gastric juices and artificial intestinal liquids must also be evaluated. These results would confirm that transgenic and nontransgenic crops are essentially equivalent, and that the transgenic plants could be utilized for food. By September 2000, the safety of 29 genetically modified crop plants for food had been confirmed (Table 1)(http://www.mhw.go.jp/).

Feed Safety Assessment

The safety of transgenic crops as feed must be approved by the Commercial Feed Division of the Livestock Industry Bureau of MAFF. Most of the evaluation items for feed safety are essentially similar to those for food safety. By December 1999, 25 genetically modified entries had been confirmed for their feed safety.

TABLE 1
Current Status of GMO Confirmed Food Safety in Japan (May, 2000)[a]

	Crop name with phenotypic category (line name)	Institution/company
1	Herbicide Tolerant Soybean (40-3-2)	Monsanto
2	Herbicide Tolerant Canola (HCN92)	AgrEvo
3	Herbicide Tolerant Canola (PGS1)	Plant Genetic Systems
4	Herbicide Tolerant Canola (GT73)	Monsanto
5	Lepidoptera Resistant Corn (Event176)	Ciba-Geigy
6	Lepidoptera Resistant Corn (Bt11)	Northrup King
7	Coleoptera Resistant Potato (New Leaf Potato)	Monsanto
8	Lepidoptera Resistant Corn (Yield Guard Corn:MON810)	Monsanto
9	Coleoptera Resistant Potato (New Leaf Potato)	Monsanto
10	Lepidoptera Resistant Cotton (Ingard Cotton)	Monsanto
11	Herbicide Tolerant Corn (T14,T25)	AgrEvo
12	Herbicide Tolerant Canola (PHY14,PHY35)	Plant Genetic Systems
13	Herbicide Tolerant Canola (PGS2)	Plant Genetic Systems
14	Herbicide Tolerant Canola (PHY36)	Plant Genetic Systems
15	Herbicide Tolerant Canola (T45)	AgrEvo
16	Herbicide Tolerant Cotton (Roundup Ready Cotton)	Monsanto
17	Herbicide Tolerant Cotton (BXN Cotton)	Calgene
18	Herbicide Tolerant Canola (MS8RF3)	Plant Genetic Systems
19	Herbicide Tolerant Canola (HCN10)	AgrEvo
20	Ripening Delayed Tomato	Calgene
21	Herbicide Tolerant Canola (MS8)	Plant Genetic Systems
22	Herbicide Tolerant Canola (RF3)	Plant Genetic Systems
23	Herbicide Tolerant Canola (WESTAR-Oxy-235)	Rhone-Poulene
24	Lepidoptera Resistant and Herbicide Tolerant Cotton (Bollgard with BXN Cotton :31807)	Calgene
25	Herbicide Tolerant Sugar Beet (T120-7)	AgrEvo
26	Herbicide Tolerant Corn (DLL25)	Dekalb
27	Lepidoptera Resistant and Herbicide Tolerant Corn (DBT418)	Dekalb
28	Herbicide Tolerant Corn (GA21)	Monsanto
29	Herbicide Tolerant Canola (PHY23)	Plant Genetic Systems

[a]Note that potato and sugar beet do not comply with the guidelines concerning environmental and feed aspects; living potato plants may not be imported from the United States because of the presence of pathogens.

Building Public Acceptance

MAFF began its public acceptance project in 1995 with a budget of US$240,000 (US$1 = ¥100). In view of the importance of building public acceptance, the budget increased by US$480,000 from this fiscal year. The purpose of this project is to provide the public with accurate information on biotechnology and genetically modified crops.

MAFF assigns responsibility for planning and executing the activities of building public acceptance to the Society for Techno-Innovation of Agriculture, Forestry and Fisheries (STAFF), an extradepartmental body of MAFF.

The Japanese government, like governments of many other countries, has to reduce personnel costs due to economic stagnation and other factors. It is difficult to justify the use of government officials to provide human resources for providing the information and training that build public acceptance. Considering these circumstances, external organizations such as STAFF are essential to the public acceptance project.

Seminars and Symposia

In collaboration with STAFF, several types of seminars were organized to provide consumers with accurate information on the nature of genetically modified crops and to establish direct relationship with consumers.

One-Week Seminars. Intensive one-week seminars were conducted for leaders of various professions, e.g., high school teachers, dieticians, prefecture government employees, private sector representatives, and farmers. The purpose of the seminars was to train people to serve as mediators between the government and consumers. Mediators should be able to explain with confidence the characteristics of genetic engineering and the merits of the technology. Therefore, they should understand the scheme of risk assessment for environmental and food safety issues, and have had exposure to laboratory work in molecular biology. From 1998, a new training course targeting high school students was started. High school students are interested in molecular biology; many are able to consider not only the safety of GMO but also global issues concerning GMO.

The seminars were held in Tsukuba in the summers from 1996 to 1999. Participants were invited through STAFF members and seven Regional Agricultural Administration Offices of MAFF. For each seminar, the organizers selected participants to maintain a balance of regional and professional interests. A total of 40 participants attended these seminars in 1999. They had an opportunity to study current developments in agricultural biotechnology and learn about risk assessment procedures regarding the environment and food safety. Participants could also attempt some molecular biology laboratory work. Because most participants had no prior experience with DNA work, they usually felt that it was mysterious or even dangerous. But they were surprised and pleased when they could isolate total DNA from rice tissue. Consequently, they experienced real situations and gained a new understanding concerning DNA work and DNA itself.

Two-Day Seminars. These seminars, which have the same basic purpose as the 1-wk seminar, are held twice each year. They are aimed especially at high school teachers and regional leaders who could not attend or were not selected for the 1-wk seminar. Participants in the 2-d seminars attend various lectures regarding the Japanese risk assessment system for genetically modified crops, and they can conduct basic experiments in molecular biotechnology. Participants report their impressions, and MAFF and STAFF use these data to improve future seminars.

"Bioforum" Sessions. Organized once every year, bioforums consist of a series of lectures and a display of products made from GMO. The purpose is to give a large number of ordinary consumers a chance to see products of biotechnology and listen to the presentations by scientists or government staff. Bioforums are announced through the Regional Agricultural Administration Office of MAFF and members of STAFF. They have been held in Tokyo and other large cities, such as Nagoya, Kyoto and Fukuoka. Bioforums will continue to be held in the same or other large cities in the future.

Informative Brochures. The seminars described above and the participants who can attend them are limited in number. Therefore, to expand opportunities for educating the public about biotechnology, MAFF produced two informative brochures. Published in August 1995, "Biotechnology in Daily Life" presents applications of biotechnology such as embryo culture, cell fusion, and genetic transformation of crops, as a wide spectrum of genetic variation. "Quick Guide to Recombinant Crops," published in September 1996, explains the basic mechanisms and merits of genetic transformation and addresses some frequently asked questions and concerns of the public. The brochures are frequently used at seminars and symposia to introduce biotechnology and improve consumers' understanding of genetic engineering. Consumers and environmental activists have shown considerable interest in the brochures. As of March 2000, ~52,000 copies of "Biotechnology in Daily Life" and 90,000 copies of "Quick Guide to Recombinant Crops" had been distributed free of charge.

Homepages on the Internet. By now, most public organizations and many private companies and home computers can access the Internet. Therefore, MAFF operates a web site on the Internet in both Japanese and English (http://www.maff.go.jp/) to provide information quickly to a worldwide audience. The Innovative Technology division has developed a web site linked to the MAFF homepage (http://ss.s.affrc.go.jp/docs/sentan). This site provides information about the use of the guidelines and the development of GMO, in addition to information related to the mission of the Innovative Technology division. Relevant new information summarized from all sections is collected by the section chief charged for public acceptance, and the site is updated about once a month. The MHW provides food safety–related information on its Web site (http://www.mhw.go.jp), as well as the brochure "Quick Guide to Recombinant Crops."

Participation in Activities of Other Organizations. As genetically modified crops become increasingly available in Japan (particularly canola, corn, and soybean), primary buyers and food-processing companies become more concerned with reactions from consumers and clients. To educate the public about genetically modified crops, companies have begun organizing meetings and symposia. Members of the Innovative Technology Division of MAFF and researchers of

National Institute in MAFF participate in symposia and panel discussions organized by consumer groups, private companies, and associations of companies to provide appropriate information. In 1999, there were 90 such participatory events.

Conclusion

By the year 2000, 15 years of field trial and 7 years of commercialization of genetically modified crops have already passed regulation in the United States and other countries including Canada and EU countries, Japan also has accumulated knowledge through many field trials. The results of many field trials of GMO indicate that no harmful effects of GMO on the environment have been reported to date (8,9). It can be concluded that risk assessments systems for the environment, food and feed have been successful.

However, some people remain uncertain about the food safety of GMO and their environmental effect. Some radical consumer and environmental activists object to cultivation of transgenic plants, consumption of genetically modified food, and research for developing novel crop plants by genetic engineering. Generally, most consumers who object to GMO may be informed incorrectly and have biased information about the risk assessment of GMO and the safety and merits of GMO. To promote an increase in public acceptance, it is important to provide accurate information and have good communication with the public. MAFF started a project for building public acceptance of GMO in 1996. In the activities of the project, MAFF organized several types of seminars and symposia, published information brochures and opened a Web site. It is important that reliable risk assessment with a solid scientific basis be implemented; such an achievement can contribute to building public acceptance for the application of GMO.

References

1. James, C., *The Global Review of Commercialized Transgenic Crops: 1999.* ISAAA Brief No. 12-1999. ISAAA, Ithaca, NY, 1999.
2. Berg, P., D. Baltimore, D.W. Boyer, S.N. Cohen, R.W. Davis, D.S. Hogness, D. Nathans, R. Roblin, J.D. Watson, S. Weissman, and N.D. Zinder, Potential Biohazards of Recombinant DNA Molecules, *Science 185:* 303 (1974).
3. Berg, P., D. Baltimore, S. Brenner, R.O. Roblin, and M.F. Singer, Summary Statement of the Asilomar Conference on Recombinant DNA Molecules. *Proc. Natl. Acad. Sci. USA 72:* 1981–1984 (1975).
4. OECD, *Safety Considerations for Biotechnology; Scale-up of Crop Plants,* OECD Publication (93 93 08 1), 1993.
5. OECD, *Safety Evaluation of Foods Derived by Modern Biotechnology: Concepts and Principles,* OECD Publication (93 93 04 1), 1993.
6. Science and Technology Agency (STA), *Guideline for Recombinant DNA (rDNA) Experiments,* STA, Tokyo, 1992, pp. 1–245

7. Ministry of Agriculture, Forestry and Fisheries (MAFF), *Guidelines for Application of Recombinant DNA Organisms in Agriculture, Forestry and Fisheries, the Food Industry and Other Related Industries,* MAFF, Tokyo, 1989, pp. 1–67.
8. *Proceedings of the 3rd International Symposium on Biosafety Results of Field Tests of Genetically Modified Plants and Microorganisms,* Monterey, CA, 1994.
9. *Proceedings of the 4th International Symposium on Biosafety Results of Field Tests of Genetically Modified Plants and Microorganisms,* Tsukuba, Japan, 1997.

Chapter 4

International Marketing of Genetically Enhanced Grains and Oilseeds: Opportunities and Challenges

Harold A. Hedley

Grains and Oilseeds Division, Market and Industry Services Branch, Agriculture and Agri-Food Canada, Ottawa, ON, Canada

Introduction

Genetically modified (GM) plants and the food produced from them have captured public attention around the world in a very short span of time. As a result of the media coverage of biotechnology and related issues over the past two years, consumers in Canada have become highly aware of foods derived from biotechnology and, in particular, from genetically modified crops. In a recent study by the Angus Reid group, exploring international awareness and perceptions of genetically modified foods, 78% of Canadian respondents indicated that they had at least some knowledge of such foods.

There are both strong advocates and strong opponents to the use of this crop-breeding technology to enhance the traits exhibited by agriculture and food crops. Opponents of biotechnology are concerned about issues such as food and environmental safety, ethics of genetic modification, labeling, consumer choice, and corporate concentration in the plant breeding and seed industries. On the other hand, some proponents of genetic modification have suggested that the results of these technologies signify the dawn of a second green revolution.

Supporters of the technology argue that genetic modification is a valuable tool with which to pursue the goal of sustainable, intensified, and diversified food production by increasing productivity, efficiency, safety, pest control, or reducing waste and losses, while taking into account the need to sustain natural resources.

The world population continues to grow—it reached 6 billion in 1999 and is expected to continue increasing for the next 50 years at which time it may stabilize; some have forecast a figure as high as 11 billion people. Human ingenuity and the first green revolution have managed to keep food production slightly ahead of population growth, but the use of new technologies will be required to sustain the increase in food production in the decades to come. Former American president Jimmy Carter, in speaking about this technology, said: "Responsible use of biotechnology is not the enemy. Starvation is. Without adequate food supplies at affordable prices we cannot expect world health or peace."

Canadian Biotechnology Strategy

The vision of the Canadian government's biotechnology strategy announced in 1998, is "to enhance the quality of life of Canadians in terms of health, safety, the environment and social and economic development by positioning Canada as a responsible world leader in biotechnology." The strategy was developed through consultations with Canadians on both the broad policy framework and on sector-specific concerns. The Canadian Biotechnology Advisory Committee (CBAC), an arms-length committee of eminent experts with diverse areas of specialization, provides advice to Federal Government ministers whose responsibilities include any aspect of biotechnology. CBAC continues to provide a forum to engage Canadians in a "national conversation" on biotechnology.

Assurance of Safety of GM Crops

The Canadian approach to the regulation of GM crops was developed over several years of public consultation. It builds on existing legislation and works through institutions that are already well positioned to assess the potential benefits vs. the risks of a product derived from biotechnology. The Canadian regulatory system protects public health and ensures environmental safety and the safety of feeds from plants with novel traits. The Canadian Food Inspection Agency (CFIA) conducts environmental safety assessments of plants with novel traits and safety assessment of feeds from such plants.

Health Canada, under the Canadian Food and Drugs Act, has established a stringent process for evaluating the safety of foods derived through biotechnology (referred to as genetically modified foods, genetically engineered foods, or novel foods). Health Canada conducts a thorough safety assessment of each biotechnology-derived food before it can be sold in Canada.

The government of Canada is committed to the ongoing process of ensuring that its regulations of foods derived from biotechnology are appropriate for the state of the science and the types of food and plant products that are being developed through future research. To ensure that Canada's food supply remains one of the safest in the world, the Canadian government, in December 1999, announced that an independent expert panel would be established to examine future scientific developments in food biotechnology. The independent expert panel will also advise Health Canada, the Canadian Food Inspection Agency, and Environment Canada on the science capacity required to enable the Federal Government to continue to ensure the safety of new food products being developed through biotechnology into the 21st century.

This panel will identify the types of foods from biotechnology that could be submitted for regulatory safety reviews over the next 10 years, the science likely to be used to develop these products and any potential risks, short- or long-term, to human health, animal health, and the environment, associated with these foods. It will also be asked to assess approaches and methodologies being developed internationally to evaluate the safety of these new foods, to identify the scientific capacity that will be required to address any health concerns related to these foods, and to identify any new policies, guidelines, and regulations related to science that may be required in this area, including the need for long-term studies to ensure human health. The panel will complete its final report by the end of the summer of 2000.

Production of Crops with Genetically Modified Traits

Farmers in those countries in which genetically modified crops with agronomic traits have been commercialized, have rapidly adopted these crops. As recently as 1996, there were only 1.7 million hectares of crops with genetically modified traits around the world. By 1999, 39.9 million hectares of crops with genetically modified traits were grown. The United States accounted for 72% of the global area in 1999, followed by Argentina at 17% and Canada at 10%.

The Canadian area of crops with genetically modified traits in 1999 was ~4 million hectares (just under 10 million acres). Three grains and oilseeds crops—canola, corn and soybeans—each with substantial production of varieties with genetically modified traits, accounted for this Canadian acreage. The total 1999 area of these three crops in Canada was ~19 million acres. About 55% of Canadian canola area was planted to varieties with genetically modified traits, whereas 35% of corn and 20% of soybeans were seeded to varieties with genetically modified traits. Total Canadian production of these three crops in 1999 was ~20.7 million tonnes; ~8.6 million tonnes of this amount was genetically modified.

Genetic modification of grains and oilseeds has brought with it new challenges in the international trade arena. Before the commercialization of genetically modified crops, Canadian farmers planting registered varieties of seed could normally be assured that their crops would be accepted in Canada's export markets. Registration of the variety of seed is the final regulatory stage in Canada before the commercialization of plants with novel traits such as canola. Before reaching that stage, the environmental and feed safety of a plant with a novel trait must have been approved by the Canadian Food Inspection Agency and its food safety must have been approved by Health Canada.

Developers of crops with GM traits face the challenge of pursuing environmental, food and feed safety assessments, and obtaining approvals not only in the country in which the seed will be grown but also in every country to which the crops grown with that seed will be exported. This has increased the regulatory challenge for plant breeders, in both private companies and other institutions developing plants with genetically modified traits. Historically, a significant amount of Canadian plant breeding has been carried out by public institutions such as universities or government research centers, which have been able to deal with the Canadian regulatory system but who have little or no capability of pursuing the necessary regulatory approvals in all markets to which a crop grown in Canada may be exported.

Export Orientation

The Canadian grain and oilseed sector is very dependent on the export market. On average, 50% of the grains and oilseeds produced in Canada are sold in the export market. This includes exports as processed products, such as vegetable oil and oilseed meal. In the last complete crop year for which data are available, 1998-1999, Canadian exports of these three crops totalled 5.3 million tonnes. In the case of canola, further exports (calendar year 1999), as oil and meal, totaled 547 thousand tonnes and 1.15 million tonnes, respectively.

International Challenges

A common element in the various trade and market access difficulties that have arisen in relation to GM agri-food products is the problem of consumer and market acceptance. In the Angus Reid study exploring international awareness and perceptions of genetically modified food referred to earlier, two thirds of respondents in all countries indicated that they would be less likely to buy a food product if they knew it contained GM ingredients. This ranged from a high of 82% of respondents in Germany to a low of 57% of respondents in the United States.

For a variety of reasons, consumers have expressed concern about the use of this technology in the agriculture industry. Government responses to this public uncertainty have varied widely from initiatives to label GM foods to bans on their domestic production or import. Management of the international trade of GM crops and their products in response to such developments will be a major challenge for all industry stakeholders.

In Canada, the Canadian General Standards Board (CGSB) at the request of the Canadian Council of Grocery Distributors has formed a standards committee on voluntary labeling of foods obtained or not obtained through genetic modification. This committee, which includes >80 stakeholders, is developing a standard for the voluntary labeling of foods derived from biotechnology. All stakeholders, including consumers, are working together to ensure that the proposed labeling standard will be both meaningful and enforceable for foods obtained and not obtained through genetic modification. Canada believes that this approach can lead to a standard for voluntary labeling of such foods that will be consistent with our international trade obligations. A principle that the committee is following is that voluntary GM food labels must be informative, clear, concise, meaningful, and verifiable.

Access for GM crops is being denied because approval systems are not working in some countries. In response to public concern about GM foods, the European Union (EU) approval system for new GM products has been effectively "suspended." No GM plants have been approved in the EU since 1998. As a result, all Canadian canola has been shut out of the EU marketplace because of the commingling that occurs in the Canadian grain handling and transportation system.

Other governments have responded to consumer concerns by looking at mandatory labeling. The specific details and requirements of the mandatory labeling regimens that are currently being considered by a number of countries vary widely. One variable is the tolerance for GMO. The EU proposal for a 1% tolerance level for the unintended contamination of non-GM foods by GM foods means that all products with >1% GMO content must be labeled as a GM product. Another challenge is the possibility that some countries will enforce a zero import tolerance for shipments that contain even a trace of GM products not yet approved in the importing country.

Mandatory labeling in key markets for Canadian exports of crops containing GMO whose foodstuffs are required to be labeled will have significant implications for the western Canadian grain handling and transportation system. Canada may have no choice but to launch an identity-preserved (IP) and segregation system for bulk grains and oilseed commodities exported from western Canada, even though this will significantly increase handling and transportation costs. At present, the majority of GM canola in western Canada crops is substantially equivalent to conventional canola, and as you know, has not been segregated through the handling and transportation system. Any segregation has been to respond to a specific niche market demand. However, the capability for segregation will be essential for the next generation of GM crops expected in the future, which may not be substantially equivalent to conventional crops. In eastern Canada, however, some GM soybeans and corn are being segregated because the different structure of the handling system in that area of the country is able to accommodate it at an affordable cost.

A segregation and IP system for non-GM and GM commodities will have to be implemented within the context of Canada's quality assurance system for grain and oilseed exports. The Canadian Grain Commission's recently released position on genetically modified grain is oriented toward providing services to farmers and the grain industry to maintain the integrity of the quality assurance system in Canada.

The necessary operation, monitoring and certification of a global system for separate handling of non-GM grains and oilseeds, will be expensive. One study estimated the incremental costs for a separate handling system for non-GM canola at up to $40/tonne. At current prices for canola, that amounts to an increase in export value of ~15%. It remains to be seen whether processors, distributors, food retailers and, ultimately, consumers will be willing to pay a premium for non-GM certified food products sufficient to cover those costs.

International Opportunities

Although there are some niche markets developing for certified non-GM commodities, this is still fairly small in the overall context of world grain and oilseed trade. The Canadian soybean industry has been able to develop significant exports of non-GM soybeans both to European countries and to Asian markets.

The Canadian grain commission is now providing certification services to exporters of non-GM soybeans. This ability to export certified non-GM soybeans has created new export opportunities for Canadian soybeans. Significant markets for identity-preserved non-GM soybeans from Canada have developed. However, very few (if any) Canadian companies are offering significant premiums to producers for non-GM crops because most foreign buyers are reluctant to pay a premium price for non-GM crops.

Producers in North America are not alone in recognizing the agronomic and environmental benefits of genetically modified crops. However, producers in many countries are overwhelmed by the challenge of sorting through all of the information on genetically modified crops and their products in order to separate the "grain from the chaff." This presents tremendous opportunities for producers from different countries to work with one another to share balanced, factual information and to utilize that information to earn and enhance public confidence in genetically modified crops.

International Forums Addressing These Issues

One message from the global biotechnology debate that is abundantly clear is the following: "The world wants a clear set of international rules in reference to biotechnology whether you are a supporter of biotechnology or not." There are a large number of parties who have a considerable stake in the rules of trade that will govern GMO. For example, farmers face uncertainty regarding the advisability of planting GM crops. Grain handling and transportation companies and food processing firms may be faced with investment for separate facilities for GM and non-GM products.

Canada is working to encourage harmonization of international standards and rules for the production, sale, and trade of agricultural commodities and food products containing or produced from genetically modified crops. Canada is pursuing this goal through several international forums and institutions such as the World Trade Organization (WTO), Codex Alimentarius, the Organization for Economic Cooperation and Development (OECD), the Food and Agriculture Organization of the United Nations (FAO), and the United Nations Environment Program (UNEP).

Public Perception and Market Acceptance

The failure or success of these GM plants and their products in the marketplace revolves around the principle of public trust. This trust must be gained by maintaining public confidence in the integrity and thoroughness of the regulatory authorities in ensuring human, animal, and environmental safety. Additionally, the public must be confident that they have been provided with the information required to understand the principles upon which those GM products and their regulation are based.

Consumer and customer preference is a significant factor that must be addressed in the context of attaining acceptance for GM crops in both international and domestic markets. An often quoted truism is that the customer is always right regardless of how the customer comes to his/her decision.

Communication to Enhance Public Acceptance

As a consequence of not having received the general information required to mitigate their fears, a significant number of consumers are basing their ideas about biotechnology and GM foods on either poorly explained and poorly understood information or even in some instances, misinformation. A critical challenge faced not only by governments, but also by industry, is gaining and maintaining public confidence in the regulation of products developed with this technology.

The agricultural industry in Canada and around the world may be at a crossroads with respect to achieving commercialization of the potential benefits of biotechnology. Effective public communication of the knowledge about these products and their regulation is critical in attaining public acceptance of them. The combined efforts of all stakeholders are required to continue to respond to the public's needs for balanced and factual information. The public, as consumers, will determine the commercial success of products developed through biotechnology. Continued consultation, communication, and partnership building are critical to earning and enhancing public confidence and present a challenge for all stakeholders.

Chapter 5

Risks and Challenges for the European Crushing Industry

Pascal Cogels

EC Seed Crushers' and Oil Processors' Federation, Brussels, Belgium

Introduction

Although it is difficult to be original on a subject so widely reported and debated worldwide, the challenges and opportunities that genetically modified organisms (GMO) create for our industry can be viewed from several perspectives. This presentation will raise and discuss a number of questions that are essential to consider in understanding the European sensitivities and approaches to biotechnologies.

Are Biotechnologies Good or Bad for Our Society?

This is not a typical European question, nor is it a candid one. It is nothing less than one of the most complicated questions to answer, i.e., good or bad from which point of view, for whom, for which part of the world, and based on what criteria? The real global debate has just begun. It is crucial that all actors in society are involved. Beyond pure economical and scientific considerations, ethical values, culture and other criteria must be covered as well; perhaps it is because, some of those legitimate values were not considered in the early stages that resistance to GMO has developed.

Some crucial questions will predominate:

- What is equivalence?
- Have we tested our novel products sufficiently?
- What are the real long-term effects of new biotechnologies on the environment(s)?Are they a threat to biodiversity?
- What will be the effect of biotechnologies on traditional agricultural practices?
- Can we harmonize our approaches worldwide and create a real global platform for the development of biotechnologies?
- How should we address the issues of responsibility and liability?
- How can we prevent trade distortions or new barriers to trade?
- How can we accommodate the systems and approaches to avoid discriminating among operators or among countries?

- Can we agree on the precautionary principle?
- Can we accept the private ownership of genomes and seeds?

The Food and Agriculture Organization (FAO), the CODEX, and the World Trade Organization (WTO) are now increasingly involved in a global debate concerning those questions. While considering all of these issues, however, we should not loose sight of the incredible benefits that can be generated by modern biotechnologies. It will be up to us to manage such progress in the right way. In the following paper, some of the European views are addressed specifically.

Are Europeans Anti-GMO?

Some might argue that European resistance represents a way of protecting themselves from competitors (mainly from U.S. agricultural competition). This is an entirely erroneous view, even though it has been promulgated in the U.S. press and in conference reports. According to a highly placed United States Department of Agriculture (USDA) official, this whole thing looked like a plot organized by EU farmers to escape the obligations resulting from the Blair House Agreement. The scenario suggested was the following: Because the United States will not be able to supply GM-free soybeans in sufficient quantities, the only alternative for the EU is to expand its own production beyond the Blair House limits. Such assertions are extremely counterproductive. They help destroy the good climate so badly needed for mutual understanding in this very complex debate about biotechnologies.

It is not to protect themselves from foreign competition that UK and EU farmers have spent billions of dollars slaughtering and burning their cattle or that the Belgian economy has lost several billion dollars throwing away food; those were their respective responses to the mad cow crisis and later the dioxin crisis. The severity with which the EU Commission is now drafting its new feed legislation is a direct response to European problems.[1] There is no deliberate intention to create new barriers.

Europe is certainly to be blamed for having been a very fertile ground for an entire catalog of food scares and scandals. This has not created the appropriate background for the introduction of new genetically modified products. Europe was also ill informed and badly organized to deal with these biotechnologies, with approval processes, with new labeling legislation, with risk management, i.e, until very recently, our food safety system has been quite archaic.

In this framework, companies have introduced new GM products without really taking into account this specific background, without undertaking a market survey, and without even consulting with customers or consumers.

---

[1]The recent declarations by Minister Glavany (France) are quite illustrative; he went so far as to say that if Europeans do not manage to harmonize their practices on the use of animal feeding material, a choice could be made to ban them altogether, to the profit of imported protein meals.

- Not only are EU food cultures and consumer attitudes different from those in the United States; even among the different EU Member States, there are wide differences in perceptions and attitudes. Scandinavia and Austria are extremely sensitive; Germany, Holland, and Belgium are more pragmatic; the UK, initially supportive, has fundamentally changed its attitude; resistance is building in France; the Southern countries of Spain and Portugal are not too worried, but Italy has begun to wonder.
- The level of trust in authorities is also much lower in the EU.
- A fragmented Europe and changing attitudes of Member States authorities have confused the whole debate.
- Contradictory statements by scientists and hints that even at the Food and Drug Administration (FDA) level, decisions might have been influenced have led the EU society to become increasingly worried. The fact that an increasing number of U.S. food companies have begun to promote segregation or have turned their backs on GM products has not helped.
- Last, but not least, pressure groups and politicians were quick to embrace such emotional subjects, sometimes to promote their own visibility, sometimes, however, in full sincerity.

Why Were EU Farmers Not More Supportive vis-à-vis GMO?

Biotechnologies are viewed by the average EU farming community as a threat for the following reasons: (i) farmers will be transformed into employees; (ii) they will no longer be able to replant their own production; (iii) they will become prisoners of complicated and binding contracts with multinationals (indeed, they might occasionally be more pragmatic about it under a different set of conditions); (iv) they are the carriers of a long tradition of good food, traditional values, and quality, whereas biotechnologies are viewed as creating uniformity, lack of taste, low quality, and potential risks.

What is very important to understand is that somehow EU farmers are not as business minded as their U.S. cousins. What is the actual value of a few percentage points improvement in yield? Certainly not the loss of freedom. After all, because they are receiving up to 60% of their total revenue under the form of hectare aids, why should they bother? That is not a good scenario for biotechnologies because, had farmers supported such developments from the outset, history would have been different!

Today, in fact, non-GM or traditional products are becoming in a way more fashionable than before and marketing campaigns have begun to promote them very strongly. Traditional, identity-preserved crops receive a premium price. Is it possible that one day such premiums will exceed the net economic advantage of planting GM crops? That remains an open question and also raises the question of the direction Europe should take because a diminishing number of countries are capable of supplying traditional beans and seeds in commodity quantities. Should Europe specialize in those crops? That also remains an open question.

What Has Been the Attitude of the EU Commission?

In the beginning, the Commission demonstrated a very supportive attitude toward biotechnology in general and toward the first GM soybeans and corn in particular (today it is likely that such products would not be authorized that easily). The EU supported the view that these products were equivalent and did not propose to label them (also the initial position of some companies). The European Parliament, however, resisted fiercely and forced the EU Commission to revisit its copy and to compromise on issues such as equivalence, labeling, origin, and identity preservation. Some Member States refused to authorize some of the products approved at the EU level (Austria, Luxembourg, and France).

The Commission brought those countries to Court, a time-consuming process. Very recently, the European Court of Justice condemned the French Government for refusing to authorize the Novartis BT Maize on its territory. Another fight took place over the issue of the definition of a threshold for adventitious contamination. It was finally set at 1%.

A moratorium was put on new authorizations, which was not lifted by the EU Commission until very recently. In the spring of 2000, however, it managed to have the EU Parliament endorse its revision of the novel food approval system and to have it reject a proposal aimed at creating a specific liability system for GMO food producers.

Are We Moving in the Right Direction? Will the EU Food Safety Authority Help Improve Things?

When the EU Commission first announced its intention to create an EU Food Safety Authority, there was much excitement and anticipation. Soon we will have a central body that will have a real say on authorizations (based on the advice of an independent panel of high-level scientists), a body that will have the authority to force Member States to apply the internal market principles. Soon the Food Safety Agency would become a credible body able to have discussions with the U.S. FDA on an equal footing. Consumer trust could gradually be restored.

Regrettably, a few weeks after this announcement we sadly witnessed the refusal of the Member State to give such a central body any real authority. They will keep (or create) their own national food safety authorities. The EU body will not have a final say on authorizations, and risk management will remain in the hands of Member States.

Why Did it Happen Like That? It is clear that Member States are fearing the interference of such a central agency in their national "affairs." Members of the European Parliament are not prepared to abandon their new powers so recently gained through the adoption of the codecision procedure.[2] Too strong a Food Safety Agency could erode their influence.

[2]Such a procedure provides for a joint decision by the European Parliament/Council of Ministers on specific matters.

How then can we envisage a friendly relationship among 25 national food safety agencies (once enlargement of the EU is completed), given the way things are presently happening with just 15 members? Preserving the triangular power structure of the EU has its price: it might well take 10 years before our dreams become reality.

However, there are some grounds for optimism: (i) the issue is not over; (ii) the EU Commission is now in the process of consolidating all EU safety issues in the hands of a single directorate called the "Directorate General SANCO," i.e., there is a clear determination to clean up the European scene; and (iii) the recent victories of the EU Commission in Parliament on issues such as the approval process and the liability question are indicative of a certain progress.

It is also my belief that the authority of bodies such as the CODEX and the Organization of Economic Cooperation of Development (OECD) will force the acceleration of new initiatives at the EU level.[3] But it is also my conviction that the role of the WTO should be confined for the time being. Biotechnologies should not be on the agenda of current trade negotiations. They should be dealt with separately. In the future, WTO dispute settlements will have to be based on principles decided at CODEX and OECD levels, in a consensual way.

A number of companies have strongly encouraged the European Commission in the creation of such an EU Food Authority. But if such a Food Authority is not given any real say at the EU level and fails to be recognized as an independent, trustworthy, scientific-based and strong body, the objectives pursued might be missed.

Will Europeans Use the Precautionary Principle as a Trade Barrier?

Again we should be careful about suspicious and aggressive attitudes; they do not facilitate the debate. There is no deliberate intention or hidden strategy developed by the EU Commission. One must understand that food safety issues have become a real concern for our citizens; our politicians have learned recently, and sometimes bitterly, not to be exposed to accusations of irresponsibility. The main principles that will be used in activating such a precautionary principle have been made public recently by the EU Commission. Nevertheless, we should not be too naive because there are those who will be tempted to take advantage of our belief. We must therefore clearly define and limit the use of such a precautionary principle and, at the same time, not question the sovereign right of countries to protect themselves should sufficient scientific evidence be lacking.

What Are the Challenges for the EU Oilseed Crushing and Refining Industry?

Our situation is far from being a comfortable one. The main problems can be defined in terms of supply disruptions, market anticipations, sourcings, costs, legal security, and harmonization at the EU level.

---

[3]Those bodies are working hard to create a global platform for discussion and evaluation of biotechnologies as well as to centralize risk assessment.

Supply Disruptions. Our cargoes have been blocked quite regularly by anti-GMO groups. This has created havoc and penalized our industry.

Market Anticipations. Distributors are placing significant pressure on us. They themselves and the food industries are targeted regularly by such groups. Understandably, those distributors and industries are changing their views about supplies, i.e., some are now requesting certified GMO-free products, some for a certificate of non-GMO origin, and others are imposing a level of threshold below the one recently adopted by EU authorities.

Under such conditions, it is quite difficult to anticipate demand. This is certainly one of the crucial issues confronted by our industry. We sense that demand for non-GM products is growing. But it is extremely difficult to anticipate by how much, what the final non-GM market share will be, whether it will stabilize or whether this trend will be reversed at some point in time. Similarly, the organic food market is growing at a very high rate; here too, however, we have some difficulties in measuring its final market size and the effect that such growth will have on our industry. The difficulty in anticipating demand has definitely increased the financial and operational risks of our companies.

Sourcings. Sourcing supplies has become a significant problem for the following reasons: (i) GMO free sourcing does not exist. (ii) Some traditional customers are now at times sourcing their own supplies (e.g., Carrefour in Brazil). (iii) Sourcing nonidentity-preserved GMO beans is not always possible beyond certain quantities. (iv) Canada, a traditional supplier of canola, has ceased to supply the EU because of admixture of approved and nonapproved varieties. (iv) Export of U.S. soybeans to the EU was almost threatened recently by the intention of a company to commercialize a new variety not yet authorized in the EU.

The Cost of Segregation and Identity Preservation

Most of the large crushing and refining companies are now engaged, to various degrees, in non-GMO sourcing and identity preservation. As time goes by, the reality seems to be changing—once a taboo, segregation is now possible. However, that does not mean that it is simple or cheap to segregate or preserve identity, which is precisely the source of the problem. There is a cost that the industry must recover. But customers are resisting the idea of paying premium prices for traditional products (in a way, paying a premium for traditional products is not exactly good publicity for GM-modified products either).

The willingness to pay a premium varies among different products, i.e., premiums are accepted for protein concentrates, lecithin, and ingredients for infant foods, but they are resisted for oils and meals. This situation may change, however, with the growing resistance to GMO foods. Ironically, new biotechnologies were the originators of a new area of competition among companies that now fight to set up the most efficient and cheapest segregation and identity preservation systems.

Another issue concerns whether to harmonize the techniques of segregation or identity preservation. The debate has just begun, and at least three different EU bodies are working on identity preservation, analyzing its costs, and attempting to define the criteria. The different questions that are likely to be raised include the following:

- How can such rules be enforced internationally (when it is already complicated to have EU Member States apply the EU decisions)?
- How can compliance with such rules be verified?
- Should companies be left to compete freely on the front of segregation and identity preservation?
- Will all operators be in a position to segregate and preserve identity?
- Will this issue create new discriminations among countries, regions, and operators?

A Lack of Legal Security

We will not debate the issue of authorizations here. The main problem today pertains to the complex, contradictory, incoherent, and unfinished EU labeling legislation. Its principles can be described as follows: (i) If raw materials used to make an ingredient are of GM origin, then such ingredients must be labeled unless it can be proved that there are no traces of modified protein or DNA in the final product. (ii) If raw materials are of a non-GM origin (certified), then they do not have to be labeled if modified traces of protein or DNA do not exceed 1%; if they exceed 1%, they must be labeled (this 1% threshold cannot be exceeded even if several ingredients are used in a product).

The main problems include the following:

- To date, no analytical system has been identified that is quantitative and sufficiently precise. Consequently, as far as refined vegetable oils are concerned, the labeling legislation is not applicable.
- It is not coherent to have a threshold for non-GM products and no threshold for GM-products.
- How can one determine that the declaration of origin is not false in the absence of clear, enforceable, controllable rules, harmonized at the international level?
- How does one ensure that minimum standards of segregation and identity preservation have been respected?

A Lack of Harmonization

Other problems arise from the nonharmonized implementation of such labeling rules by the Member State authorities or by the lack of respect for EU decisions by the same Member States (France, Luxembourg, and Austria). This is forcing companies to adapt to a number of legal environments, thus increasing marketing costs.

Conclusions

Problems have thus far outweighed opportunities. We are eagerly awaiting the second generation of genetically modified products, one that will offer consumers direct and visible advantages in terms of quality, taste, functionality, and possibly price. It is very likely that seed companies will have to pay a great deal of attention to the way in which they relate to farmers worldwide. Fediol recently listed some of the conditions it thinks are necessary to overcome EU consumer resistance: (i) the creation of a recognized and respected EU Food Authority; (ii) a transparent and efficient process of approval for novel food products; (iii) a free choice between GM and non-GM products; (iv) respect for rules across the EU; (v) respect for the principles of free circulation of approved products throughout the EU; (vi) the implementation of a reliable and objective system of traceability; (vii) a reliable system of certification of origin; and (viii) equal treatment of imported products vis-à-vis EU-produced products.

Our role is to survive and to respond to consumer demand. There is no alternative. The choice between GM and non-GM products should be offered to consumers, but without discrimination. Consumers must know what they are buying. The market should decide what it wants but at the same time, we must create the conditions for the safe development and acceptance of biotechnologies.

Chapter 6

Leading USDA Research into the 21st Century

Floyd Horn

USDA, REE, Agricultural Research Service, Washington, D.C.

Introduction

At the end of the 19th century, an American scientist, named George Washington Carver, presented the world with some important ideas. Briefly, Carver believed that there was no such thing as plant waste material and that agricultural crops had other uses besides food. Carver also believed that all people, regardless of their differences, could work together. This symposium on genetically modified crops, attended by varied segments of the global oilseed industry, is an important opportunity to do just that.

As oil chemists, you know that both traditional plant breeding and biotechnology have transformed plant crops into thousands of industrial products in addition to their vitally important uses for food, feed, fiber, and ornamentals. Today, biotechnology is providing us with an expanding list of potential biobased products and bioenergy. The USDA's National Plant Germplasm System (NPGS), with its more than 430,000 germplasm samples, has served as a key global resource for the genes that make these improvements possible. The NPGS will play an even more important role in providing genes for these new traits.

Traditional Plant Breeding and Biotechnology

Agricultural biotechnology is a tool, i.e., the application of modern scientific techniques to modify and improve crops. The ability to introduce new genes into plants from very distantly related species was not possible until agricultural biotechnology came into being. Historically, plant breeders improved crops for traits such as increased yield and pest resistance using only traditional plant breeding. Traditional breeding is still important, and it has a place in the future alongside biotechnology. In fact, agricultural biotechnology relies on traditional breeding to place most new traits into improved varieties. A role that USDA will play with these expanding uses is the development of research tools, such as molecular genetic maps and genomic databases, that enable more rapid development of new varieties, whether by traditional methods or modern biotechnology.

Traditional crop breeding has delivered the goods, food, fiber, and ornamentals, in great abundance for the United States and for the world. In 1940, one U.S. farmer fed 19 people. Today one U.S. farmer can feed 129 people. Our global population is growing, and available farmland is shrinking. Some population forecasters predict a world population of 9 billion people by the year 2050. Our ability to feed and meet the consumer needs of this growing population will keep both plant breeders and biotechnology scientists in demand.

The Promise of Biotechnology

We have now entered the 21st century, with molecular tools and new genomic information about microbes, plants, animals, and humans. Biotechnology gives us a new ability to gain even more knowledge about all living organisms in our world. It also allows us to be more specific about the genes we transfer. Continued use of biotechnology will depend on how we resolve our differences, differences that arise not only in scientific debates but also in the social, political, and ethical realms.

The promises of biotechnology includes more nutritious foods, a cleaner environment, along with better medicines and vaccines to save lives and help people live healthier and longer lives. With biotechnology, microbes, plants, and animals can be made to be more productive, grow faster, or require fewer pesticides, fertilizers, water, and other inputs.

Agricultural biotechnology has already changed the way farmers grow their crops. Most farmers now grow high-yielding crop varieties to produce standard commodities for which the specific end use may not be known. The end use, whether it is animal feed or corn flakes, is not known until the commodity reaches the processing stage. Now, farmers are beginning to grow specialty crops of high value tailored to specific end uses, i.e., they are known before the seed is sown.

Today, nearly one-half of the U.S. production acreage for corn, soybeans, and cotton is planted in genetically engineered plant varieties. These new varieties contain genes for herbicide resistance in soybeans and insect resistance in corn or cotton. As research and development continues, yet other genetically engineered commodities await commercialization.

Nevertheless, applications of biotechnology in agriculture are just beginning. For example, currently available biotechnology plant varieties are modified for only a single gene. Soon, scientists will be able to incorporate multiple genes at a time. This has already been achieved in a new rice variety, modified for high levels of β-carotene.

The first U.S. biotechnology plant varieties were commercialized only ~4 years ago, but that culminated more than 20 years of research and development. Initially, U.S. farmers gave biotechnology crops an enthusiastic thumbs up. By planting the new varieties, farmers demonstrated their high expectations of biotechnology to revolutionize agriculture. More recently, these expectations are being brought up short by the reaction of our European trading partners and other countries.

Outlining the Problems

For various reasons, not all of which are science based, some people in Europe, Japan, and even in the United States object to the use of biotechnology in agriculture.

In contrast, people in some developing nations, especially China, have embraced genetically engineered crops. Reluctant to accept agricultural biotechnology, Europeans and others are enthusiastic about medical advances brought by biotechnology. Pharmaceutical biotechnology has developed new products and therapies not available through other means. New medicines and vaccines have already been developed and may provide people more effective treatments for devastating diseases, such as Parkinson's and Alzheimer's, diabetes, heart disease, AIDS, and cancer.

If agricultural biotechnology is to be more widely accepted at some future date, it will likely be when second and third generation products are realized, offering immediate consumer benefits that cannot be achieved by conventional means.

In January, 2000, representatives of 130 countries met to reach an agreement called the Montreal Protocol on Biosafety. Both advocates and opponents of biotechnology claimed victory from these negotiations. Advocates are encouraged because the treaty gives importance to basing trade decisions on sound science. On the other side, opponents are satisfied because the treaty permits countries to refuse imported genetically modified organisms (GMO) commodities based on the "precautionary principle."

The "precautionary principle" is based on the idea that not enough is known about the long-term effects of GMO on human health or the environment This principle will be difficult to satisfy because it is nearly impossible to alleviate every possible concern about biotechnology crops before they are grown and used. In spite of its ambiguity, the Montreal Protocol has strengthened dialogue and scientific exchange of information between governments to evaluate both the benefits and risks of biotechnology.

Regulation of Biotechnology

Risk assessment of genetically engineered crops has not been overlooked in the United States. Our regulatory system for releasing and field testing biotechnology crops has been in place since 1986. The USDA is one of three Federal agencies that regulate biotechnology. The other two agencies are the Environmental Protection Agency (EPA) and the U.S. Food and Drug Administration (FDA). GMO products are regulated according to their intended use, and some products are regulated by more than one Federal agency.

In 1989, the National Academy of Science (NAS) published "Field Testing of Genetically Modified Organisms: Framework for Decisions." This publication provides guidance for U.S. regulatory agencies. Last December, NAS announced that it will update its review of the scientific process used by USDA's Animal and Plant Health Inspection Service (APHIS).

The Agricultural Research Service (ARS) is the USDA's chief scientific research agency. Our research scientists provide support for regulatory agencies, such as APHIS, FDA, and EPA. APHIS evaluates products for their potential risk to other plants and animals; the FDA reviews biotechnology's effects on food safety; and the EPA examines any products that can be classified as pesticides, including plant insect resistance. Our U.S. regulatory system is working to ensure safety for plants, animals, humans, and the environment. But it is not infallible. It can and should be improved where and when possible.

Some people fear that GM crops might introduce genes into nonallergenic foods and trigger allergic reactions in people. In the United States, makers of biotechnology foods screen their products for potential allergens, and they act quickly to protect their customers. As an example, let me cite the case of a soybean variety genetically modified to express a Brazil nut protein. When its potential for adverse effects in people with a sensitivity to Brazil nuts was discovered, the company immediately and voluntarily discontinued production before commercialization.

ARS research studies are underway to assess the effect of gene flow from biotechnology crops to weedy relatives or non-GMO crops that may be planted nearby. For example, ARS and University of Wisconsin researchers are evaluating the genetic effect of commercial transgenic squash and potato plants on their cross-compatibility with their wild and weedy relatives. This work will answer critical questions about gene flow from biotechnology crops to plants and weeds and will provide a scientific model for risk assessment.

The effect of GMO crops on nontarget species, such as Monarch caterpillars, also demands more scientific research data. As you know, controversy exploded last summer resulting from a preliminary laboratory study by Cornell University researchers. These researchers found that Monarch caterpillars died after eating milkweed heavily saturated with Bt corn pollen in the laboratory.

Unfortunately, much public distrust of GM crops was provoked by media attention to the Cornell study. Soon thereafter, ARS and Iowa State University researchers in Ames, Iowa, conducted a study, one of the first to assess the effect of Bt corn pollen on caterpillars in the field and to study pollen deposition in and around corn fields. Preliminary data strongly indicate that pollen from the majority of commercially available Bt corn varieties had little, if any effect, on Monarch caterpillars, with one exception. One commercially available transgenic Bt corn line was toxic to Monarch caterpillars, but we can reasonably expect that limited exposure to pollen might reduce any negative effects. EPA has always assumed that Bt pollen would affect all moths and butterflies, but the risk of exposure is considered low to nonexistent. Because corn pollen is heavy, it is deposited mainly within 10–15 feet of the cornfield. Our study showed that only 30% of the Bt pollen was deposited on milkweed leaves, and 90% of that was washed away when it rained. Dr. Mark Sears of the University of Guelph, Ontario, confirmed the results of the ARS-Iowa State study. Pollen deposition studies were confirmed by University of Maryland and University of Nebraska researchers. GMO opponents have suggested that these research studies were supported by industry, which is simply not true! The ARS research evaluating the effects of transgenic corn on nontarget insects was funded in-house. More research will be conducted to determine the extent that monarch adults lay eggs on milkweed plants in cornfields.

Risk Assessment

As indicated here, ARS is committed to studying all aspects of genetically modified crops, including risk assessment. We are reviewing our research program for any experiments that might cause concern about environmental risks. Where appropriate, we will conduct additional research to gain a better understanding of the potential risk.

Although crop productivity is expanding, we also must recognize that the channels for supply of seed and fertilizers are being consolidated rapidly into only a few multinational companies. The rights to much of modern biotechnology are owned and controlled by these large companies. This scenario has created unease among some farmers and processors.

Perhaps it was this backdrop that caused so much controversy regarding the development of a way to control and target gene expression in cotton seeds. The media dubbed it the "terminator technology," but its technical title is the Technology Protection System. This pioneering research by ARS scientists and Delta and Pine Land Company, a major cotton seed supplier in the United States, lays the groundwork for minimizing the risk of spreading unwanted traits into wild plant species. This technology was granted a U.S. patent last year. Its potential offers more advancements to research in developing high-value crop varieties. Yet another positive outcome could be an increase in the number of seed companies that will commit to and support further biotechnology research.

Current Status

At present, the USDA has approved ~50 GMO plant varieties. In an attempt to ensure that our regulatory system addresses controversial public issues, the Secretary of Agriculture has established an advisory committee of 38 individuals from government, academia, production agriculture, agribusiness, and environmental and consumer groups. Members of the committee include both supporters and critics of biotechnology. Some positive actions may arise from the current GMO debates. These could be better, ecologically sound ways to avoid problems. The Secretary of Agriculture has also called for the establishment of 8–12 regional pest management centers in the United States. These centers will assist in evaluating biotechnology products over a long period of time, develop and share scientific expertise, and provide information to growers, consumers, researchers, and regulatory agencies.

Meanwhile, many U.S. farmers are not deterred from using biotechnology commodities. Some seed handlers have agreed to separate those commodities from the traditional varieties. As a result, there may be increased costs associated with the segregation and testing of non-GMO commodities, but it may be necessary to ensure their future for specific markets.

The European Union has asked for mandatory labeling for any food containing more than 1% biotechnology ingredients. In the United States, we are just beginning to discuss the ramifications of labeling in our food processing system. Congress will likely consider this year whether to mandate labeling of GMO products. This brings us to recognize that food safety is inextricably linked to the continued use of agricultural biotechnology.

Behind the scenes, a less publicized organization has formed a task force with the goal of evaluating food safety. This group is Codex Alimentarius, a United Nations organization representing 165 countries. Along with the precautionary principle, this task force will consider the concept of "substantial equivalence" as established by the Organization for Economic Cooperation and Development (OECD) in 1993. Substantial equivalence offers a way to identify meaningful differences between biotechnology and traditional products.

Conclusion

Much more work is required to resolve the many differences surrounding agricultural biotechnology. Biotechnology crops can revolutionize agriculture's ability to meet the energy, nutritional, and other needs of people throughout the world.

With more improvements, biotechnology oilseed crops can supply industrial products such as lubricants, inks, resins, plastics, surface coatings, polymers, and fibers. Recognizing this need, President Clinton signed an Executive Order regarding the use of Biobased Products and Bioenergy on August 12, 1999. This Executive Order sets a goal of tripling U.S. use of biobased products and bioenergy by 2010. Meeting this goal will require a substantial commitment to agricultural biotechnology.

Our differences about biotechnology may continue in public and scientific debates for some time. What would George Washington Carver think of our expanding ability to use plants for varied purposes and to reduce waste materials? I think he would welcome it. Like Carver, we believe that all people, regardless of their differences, can work together for a bright future.

Chapter 7

Molecular Genetic Technologies for Rapid Discovery and Deployment of Naturally Occurring Plant Genes

Perry B. Cregan

Soybean Genomics and Improvement Laboratory, USDA, Agricultural Research Service, Beltsville, MD 20705–2350

Introduction

The use of marker-assisted selection (MAS) for plant improvement has its origins in the 1920s with the report by Sax (1) of genetic linkage between seed size and seed coat color in the common bean. He suggested that the linkage might be used to separate individual genes for quantitative characters. Thus, once genetic linkage is established between a readily observed genetically controlled phenotype (a visible or measurable trait of an organism) and a gene controlling a more difficult to measure trait, the phenotype can be used to select indirectly for the more difficult to measure trait. Sax (1) suggested that the form of a gene (allele) controlling a specific seed coat color in the common bean [*Phaseolus vulgaris* (L.)] could be determined to be closely associated or genetically linked with the form of a gene (allele) that produced a seed with predictably larger size. Thus, selection based upon seed coat color would allow indirect selection for altered seed size. Although such a system has great theoretical appeal, in practice, easily observable color or morphological phenotypes are rare, and discovering those that are controlled by genes linked to genes of interest is even less likely. However, with the advent of molecular genetic markers, large numbers of readily observable "molecular genetic phenotypes" could be generated with relative ease. Molecular genetic markers made it possible to seriously contemplate a generalized system of MAS for the purpose of plant improvement.

Genetic Markers

For purposes of applying genetic markers to the improvement of plants, a genetic marker is defined as a detectable difference between two or more organisms that is controlled by a single difference in the DNA of those organisms. Classical genetic markers are those variations in DNA that result in a difference in the color, size, shape, disease resistance, or quality, for example, between two individuals. The commonly used molecular genetic markers fall into two categories. Polymorphisms or differences in the form of a particular enzyme comprise the first category. The different forms of an enzyme referred to as "isozymes" are usually detected using starch gel electrophoresis. Although >40 individual genetic loci–conditioning isozyme and protein polymorphisms have been defined in soybeans (*Glycine max* [L.] Merr.) (2), this number is clearly inadequate to produce a genetic linkage map with sufficient marker saturation to detect genes dispersed throughout the 20 pairs of soybean chromosomes. In contrast to the relatively few isozyme loci at which polymorphism can be readily assayed, DNA-based polymorphisms were proposed as a rich source of molecular genetic markers. DNA restriction fragment length polymorphisms (RFLP) proposed by Botstein *et al.* (3) permitted the creation of saturated linkage maps and provided the basis of map construction in numerous mammalian, insect, and plant species.

Despite the great utility of RFLP-based molecular genetic linkage maps, substantial effort was expended in the development of alternative molecular genetic markers that detected greater levels of molecular genetic variation or whose detection was technically less demanding than RFLP. In addition, the advent of the polymerase chain reaction (PCR) (4) technique for the geometric amplification of specific DNA fragments spurred interest in the development of PCR-based DNA markers. Thus, in the late 1980s and early 1990s, PCR-based marker systems, including microsatellite or simple sequence repeat (SSR) (5–7), random amplified polymorphic DNA (RAPD) (8) or arbitrary primer-PCR (AP-PCR) (9), and amplification fragment length polymorphism (AFLP) (10) markers, began to be developed and used for genetic map construction and gene discovery.

DNA-Based Genetic Markers of Soybean

In the case of soybeans, the first genetic map based on DNA markers was published in 1990 and contained 150 RFLP markers with a total length of ~1200 recombination units (11). This map, developed jointly by the USDA, ARS, and Iowa State University with support from the American Soybean Association, is referred to as the "Public" map and saw further expansion to a total of >500 RFLP loci (12–14). The details of the public RFLP map and other similar maps and large amounts of related information can be accessed on the World Wide Web in SoyBase, the USDA, ARS Soybean Genome Database (http://129.186.26.94/).

One of the largest available AFLP maps of any plant species is available for soybeans (15). In terms of the number of marker loci, this map is second only to the USDA/Iowa State map with a total of 840 loci, 650 of which are AFLP. The first demonstrations of SSR allelic variation and heritability in a plant species occurred in soybeans (16,17). This was followed by reports of high levels of SSR allelic variation in soybean germplasm (18,19). Subsequently, 40 SSR loci were developed and mapped in soybeans along with 118 RFLP and RAPD markers (20). This study indicated that SSR loci were distributed fairly randomly in the soybean genome. Because of their high level of polymorphism, single locus nature, and random distribution in the genome, it was concluded that SSR markers would provide an excellent complement to RFLP markers for use in soybean molecular biology, genetics, and plant breeding research. The most recently released genetic linkage map of the soybean genome (21) was a composite of three different maps that in total contained in excess of 600 SSR markers as well as >500 RFLP loci. This map also contained a number of classical, isozyme, RAPD, and AFLP loci.

Single Nucleotide Polymorphisms (SNP): A New Generation of DNA Marker

Single nucleotide polymorphisms and small insertions and deletions, collectively referred to as SNPs, are by far the most abundant source of DNA polymorphisms in the organisms in which they have been studied in detail, including humans (22,23) and mice (24). In humans, these variations are estimated to occur at a frequency of ~1/1000 base pairs (bp) when any homologous chromosome pair is examined (25–27). A recent survey that examined 2.3 Mbp of human DNA discovered 1.0 SNP/kbp in a population consisting of three unrelated individuals (six chromosomes) and ~1.4 SNP/kbp among seven unrelated individuals (28). Only limited data on the frequency of SNP in plants are available. Cho *et al.* (29) compared the DNA sequence of >500 kbp of the two *Arabidopsis thaliana* genotypes, Columbia and Landsberg, and detected one SNP/1034 bp. Bhattramakki *et al.* (30) assayed 30 maize inbreds and found in excess of 14 SNP/kbp. In soybeans, Zhu *et al.* (31) estimated 1.64 and 4.85/kbp in coding and noncoding DNA of soybeans, respectively, on the basis of the sequence analysis of 50 kbp in 22 diverse cultivated soybean genotypes. This is a somewhat higher rate than reported in humans and suggests that SNPs can serve as genetic markers to complement and greatly augment existing markers for the creation of genetic linkage maps. In addition, human geneticists have suggested the use of SNPs in "association analysis" (32). Association analysis compares allelic frequencies in patients and controls to dissect the genetic control of complex genetic diseases. The genome position of genes responsible for a disease can be detected on the basis of a significant difference in the frequency of alternative alleles of a SNP or SNP haplotype between patients and controls. The power of this approach to define the genetic control of disease and to assess risks of abnormal response to drug therapy was responsible for the initiation of numerous public and private research programs to discover SNPs in the human genome. The Human Genome Project leaders were particularly concerned that human SNP discovery take place in the public sector before patents and restrictive intellectual property attachments would inhibit many researchers from using this powerful analytical tool (33). As a result of this concern, a $45 million research program was initiated to discover and catalog perhaps as many as 500,000 SNP over a 2-y period (34,35). The SNP data will be freely available to all researchers, both public and private.

Applying Molecular Genetic Markers to Plant Improvement

The availability of a wide range of high throughput systems for the detection of SNPs and the likelihood that SNP analysis will continue to improve in terms of cost per data point suggest that SNP DNA markers hold great promise for MAS in plants. The successful use of MAS for the improvement of crop plants has two basic phases, calibration and application. Calibration is the association of a particular form (allele) of a genetic marker with the form (allele) of a gene or DNA segment that produces a desired phenotype. Interest has focused on the association of genetic markers with genes controlling both qualitative and quantitative traits. The former are those traits under the control of a single gene, whereas the latter are traits whose expression is under the control of more than one gene. These quantitative trait loci (QTL) are the genes that control traits such as resistance to abiotic stresses, some grain quality traits, as well as seed and dry matter yield. An oversimplified description of the process of MAS can be briefly outlined for a self-fertilizing species such as soybeans as follows. In the calibration phase, individuals (parents) with contrasting genetically controlled characteristics or phenotypes for a trait(s) of interest are hybridized to create a population of progeny. As a result of genetic recombination, the progeny will possess combinations of alleles derived from both parents at genetic marker loci as well as at loci that control the phenotype of interest. In the case of soybeans, a naturally self-fertilizing species, populations are allowed to self-fertilize to homozygosity to create recombinant inbred lines (RIL). The development of such a population is outlined in Figure 1. In this case, a DNA marker locus "M" is located in close proximity to one of the loci (Q) controlling a quantitative trait. Because the QTL and the marker locus are in close proximity, almost all genotypes that carry the "M" allele at the marker locus carry the "Q" allele at the QTL locus. Similarly, almost all genotypes that carry the "m" allele at the marker locus carry the "q" allele at the QTL locus. If one assumes that QQ homozygotes are on average measurably superior to qq homozygotes, then the close associate between the M and Q loci dictate that MM homozygotes will similarly be measurably superior to mm homozygotes. The association of the desirable allele at the QTL with a specific allele at the marker locus is the goal of the calibration phase of MAS. The implementation of MAS involves using the marker loci identified in the calibration phase to select indirectly for desired QTL alleles. As Figure 1 demonstrates, recombination between the M and Q loci (individuals 6 and 10) reduces the effectiveness of indirect selection.

DNA Marker Technologies and Technology Transfer from the Human Genome Project

An important aspect of the "Genomics Revolution" is the fact that it is built upon technologies for the analysis and manipulation of DNA. The source of DNA is of little importance; thus, most of the same sophisticated physical and electronic analyses and manipulations that are used by human geneticists can also be used by plant geneticists. This fact provides the opportunity for plant geneticists to apply a rapidly expanding set of analytical tools to the discovery of important plant genes and to their deployment in new and improved plant varieties. Ever more powerful and cost-effective tools are at the disposal of plant breeders and geneticists. The

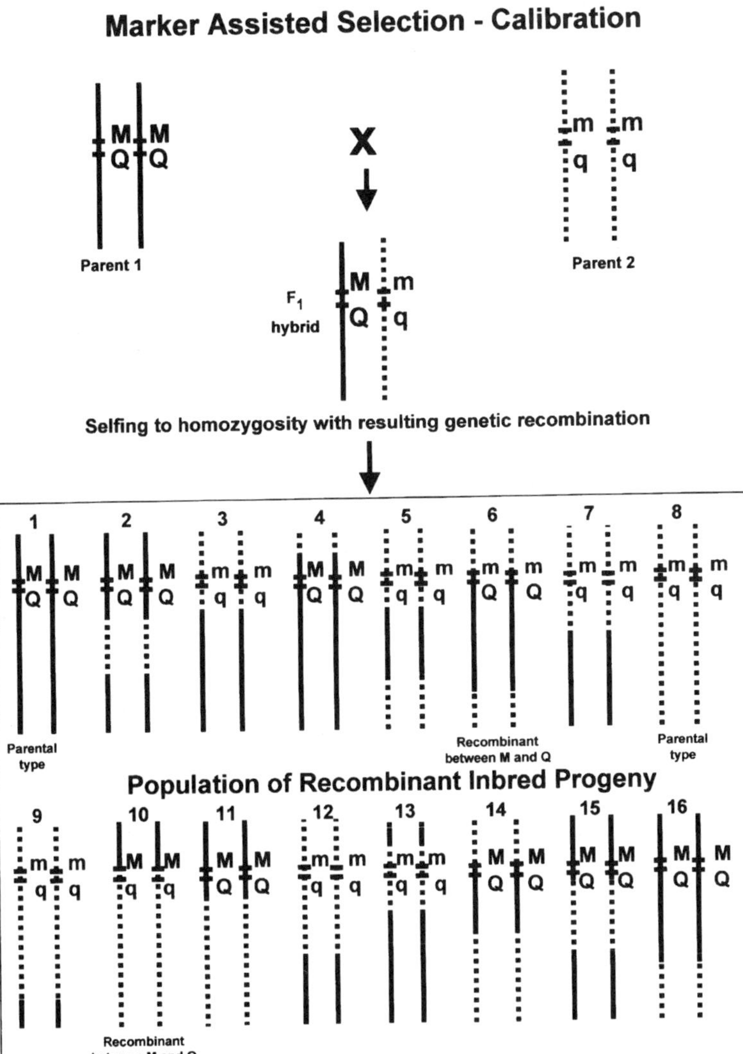

Fig. 1. A generalized scheme demonstrating recombination in one chromosome as would occur in the development of a recombinant inbred population. Two parents carrying different forms (alleles) of a QTL (Q vs. q) and a closely linked marker locus (M vs. m) were crossed to produce an F_1 hybrid that was selfed to homozygosity. The resulting population contains recombinant inbred progeny that carry DNA segments from both parents as well as a few individuals that are identical to the parents (parental types). Because of the close proximity of the M and Q loci, recombination between the two is relatively rare as indicated by the presence of only two progeny with recombination between M and Q. During the calibration phase of marker-assisted selection, the close association of the marker allele with its linked QTL allele would be determined.

following paragraphs briefly describe a few of the technologies for SSR and SNP analysis that have arisen from human genome research and that, with the availability of appropriate sets of SSR and SNP markers, will dramatically accelerate the pace of genetic improvement of soybeans and other crop plants.

SSR Allele Sizing

Slab Gel-Based Systems. Because SSR alleles are defined by the number of repeat units present in the SSR, the basis of allele sizing is generally the determination of PCR product size. As a result, numerous methods can be used for allelic discrimination. Initially, most SSR allele sizing was performed on high-resolution denaturing polyacrylamide sequencing gels to distinguish fragments that might differ in length by only 1 or 2 bases. This standard procedure was described by Cregan *et al.* (36) and is used to separate products that are either internally labeled or end-labeled with ^{32}P or ^{33}P or silver stained as described by Bassam *et al.* (37). These procedures use standard equipment that is likely to be available in most laboratories conducting plant molecular biological research.

A number of more sophisticated technologies are available for SSR allele sizing. For example, the Perkin-Elmer Applied Biosystems 377 gel-based DNA sequencer (http://www2.perkin-elmer.com/ab/about/dna/377/377a1a.html#DNAsizing) is being used for plant genotyping in soybeans (38–40). These systems employ automated sizing of fluorescent-labeled SSR alleles in which one of the PCR primers to an SSR locus is labeled with one of three different fluorescent tags. The resulting PCR product(s), carrying different colored fluorescent labels, are separated on polyacrylamide sequencing gels. A series of size standards labeled with a fourth fluorophore is employed in each gel lane to allow exact sizing of the labeled SSR alleles. Reputed advantages of the automated system compared with sizing of PCR products on standard sequencing gels are as follows: (i) the capability of multiplex analysis of several products in the same lane within the same size range, but with different fluorescent labels; (ii) single-base resolution over a wide size range from 75 to 500 bases; (iii) automated sizing; (iv) automated data output; and (v) elimination of radioactivity.

Alternative polyacrylamide gel systems are available, including the infrared automatic DNA sequencer from LI-COR (http://www.licor.com). As many as four or five SSR loci with nonoverlapping size ranges can be analyzed per lane. True images containing infrared labeled SSR fragments and standard sizing ladders (ranging from 50 to 350 bp) allow the exact sizing of SSR alleles. The ALF express II DNA Analysis System (http://www.apbiotech.com/stiboasp/showmodule.asp?nModuleid=40022) from Amersham Pharmacia Biotech also employs a single-dye chemistry format. The detection system relies on laser-induced fluorescence of the carbocyanine dye Cy5 with which SSR-containing PCR products are labeled. A horizontal ultrathin polyacrylamide gel electrophoresis apparatus is available through GeneSys Technologies (http://www-hgc.lbl.gov/instr/brumley.html). The system employs denaturing polyacrylamide gels with up to 60 lanes and can be analyzed in <90 min. Gels can be reused up to five times. Reusable gels, coupled with automated sample preparation and gel loading, enable the system to run unattended for extended periods.

Capillary Electrophoresis. The same fluorescent chemistry employed in the Perkin-Elmer Applied Biosystems gel-based systems described above is used in a single capillary electrophoresis system available from the same company (http://www2.perkin-elmer.com/ab/about/dna/310/). Automated 16-capillary (http://www.appliedbiosystems.com/3100/index.html) and 96-capillary systems (http://www2.perkin-elmer.com/ga/3700/index.html) are available if greater sample throughput is desired. These last two systems promise more rapid accumulation of SSR genotyping data as a result of concurrent multiplex analysis of SSR allele sizes in multiple capillaries.

Strand Displacement Amplification (SDA) on Microchip Arrays. Active microelectronic DNA arrays for SSR analysis employ electronic-based hybridization to provide speed and sensitivity (41,42) as developed by Nanogen (http://www.nanogen.com/Products/SDA.asp)

Mass Spectrometry. Braun *et al.* (43) proposed the use of matrix-assisted laser desorption/ionization time-of-flight (MALDI-TOF) mass spectrometry for SSR allele sizing. This system requires the annealing of a single detection primer within a few bases of the 3′-end of the SSR. A DNA polymerase extends this primer through the SSR by primer-directed DNA synthesis. A ddNTP is included to terminate the reaction at a point past the 5′-end of the SSR. Extension reactions from different SSR alleles yield products that differ in length by the number of bases in the alleles. These products are resolved using MALDI-TOF mass spectrometry.

Single Nucleotide Polymorphism (SNP) Detection

Many new technologies for SNP detection are available or are in the process of development. These technologies are based on a few general SNP detection approaches including single-base primer extension assays, allele specific hybridization or allele specific oligonucleotide hybridization, the oligonucleotide ligation assay, and the 5′ nuclease assay. These and other approaches have been reviewed by Kwok and Chen (44) and Landegren *et al.* (45). The most notable feature of SNPs as a genetic analysis tool is the great diversity of systems that have been devised for their assay and the rapidity with which allelic discrimination can be accomplished. Most of these systems depend on one of the approaches listed above; however, there are alternative strategies for SNP detection. A listing of a few of the technologies for SNP discovery and/or detection and a source for additional information about each follows.

Oligonucleotide Arrays on Glass Slides. A general protocol for SNP detection using the allele specific hybridization approach in an oligonucleotide array was described by M. Schena and R. W. Davis, Department of Biochemistry, Beckman

Center, Stanford University and is available at (http://arrayit.com/DNA-Microarray-Protocols/#Protocol10).

The GeneChip. As many as 400,000 oligonucleotides can be synthesized *in situ* and can then be used for the parallel discovery or detection of SNPs (46,47). This technology is available through Affymetrix (http://www.affymetrix.com/technology/tech-probe.html) and is based on allele-specific hybridization.

Gel-Based and Capillary Electrophoresis. Four-color fluorescent DNA sequencing chemistry identical to that employed in the Perkin-Elmer Applied Biosystems DNA sequencers can be used in single-base extension assays and analyzed on any of the gel-based or capillary electrophoresis systems described above.

The Invader Assay. Third Wave Technologies has produced the Invader® assay to discriminate single nucleotide polymorphisms (http://twt.com/invader/invader.html).

Strand Displacement Amplification (SDA) on Microchip Arrays. Active microelectronic DNA arrays can be used for SNP as well as SSR analysis and employ SDA as developed by Nanogen (http://www.nanogen.com/technology/nanochip-workstation.asp).

5′ Nuclease Assay. The TaqMan technology from Applied Biosystems for SNP allelic discrimination uses the Applied Biosystems Sequence Detection Systems such as the ABI PRISM 7900HT (http://www.appliedbiosystems.com/molecular biology/press-releases/7900launch/).

Fiber Optic Bundles. This high-throughput SNP genotyping based on BeadArray technology combines fiber optic bundles and specially prepared beads that self-assemble into an array. Each fiber optic bundle contains randomly ordered fiber optic sensor arrays (48). The technology is available from Illumina (http://illumina.com).

Mass Spectrometer-Based Systems.

- MALDI-TOF mass spectrometry for the multiplex detection of SNPs (49).
- Cleavable Mass Spectrometry Tags (CMST) for rapid SNP discovery and genotyping as proposed by Rapigene (http://www.bioportfolio.com/erbi/chiro2.htm).
- The combined use of MALD-TOF mass spectrometry and primer extension chemistry with high-density SpectroCHIP arrays for SNP genotyping (50) as employed by Sequenom (http://www.sequenom.com/home/seq-flash.asp).

Flow Cytometry. Flow cytometry provides a platform for SNP detection that can accommodate single-base extension (51), the oligonucleotide ligation assay (52), or allele-specific hybridization as SNP detection approaches (http://www.luminexcorp.com/).

Molecular Beacons. Molecular beacons are oligonucleotide probes that can report the presence of specific nucleic acids in solutions (53,54) (http://www.molecular-beacons.com/).

The Development of a Soybean SNP Map

More than 1000 SSR markers will be available for scanning the soybean genome for genes or QTL that control traits of importance. When marker and gene associations are detected, the markers can be used in high-throughput MAS systems to create genetically improved germplasm. In the case of SNP DNA markers, much research is required to discover and develop robust assays for thousands of SNPs spread throughout the soybean genome. These SNPs can be readily integrated into the existing SSR-based soybean genome linkage maps. SNPs can be identified at defined positions in the genome via the use of existing DNA sequence data from SSR flanking regions, RFLP probes, and bacterial chromosome ends that are already positioned on the soybean linkage map. Furthermore, the SSR or RFLP markers can be used to identify BAC clones, from which end sequence or subclone sequence can be obtained to discover SNPs at defined positions in the genome. In addition, a relatively large amount of DNA sequence data are available in GenBank and can be used as a source of data to develop PCR primers to use in resequencing for SNP discovery. Most importantly, the Public Soybean EST (expressed sequence tag) project (http://129.186.26.94/soybeanest.html) has as its goal the 5′-end sequencing of 300,000 ESTs. This will be followed by the assembly of ~30,000 unique EST or "UniGenes." The UniGenes will provide a rich source of soybean DNA sequence that can be mined for SNPs for the creation of a so-called transcript map, i.e., a genome map on which the positions of transcribed genes are defined. Most importantly, these genes will be the genes that control the appearance, pest resistance, oil and protein quality, and productivity of the soybean plant. The positioning of a large number of UniGenes on the genetic map will provide candidate genes that can be associated with mapped classical genes and QTL. These candidate genes will be the place to begin the search for naturally occurring genetic variation in the extensive soybean germplasm collection maintained by the USDA, Agricultural Research Service. It is important to the future of soybean improvement that an intensive effort be made to create a SNP-based transcript map of the soybean genome. Soybean geneticists and breeders will then be in the position to take full advantage of the myriad technologies arising from human genome research to discover and use the naturally occurring genetic variation present in our carefully preserved germplasm collections.

Aknowledgment

The author wishes to thank the United Soybean Board for the support received via research grant # 9222 that facilitated the development of soybean SSR markers and the assessment of SNP in soybean.

References

1. Sax, K., The Association of Size Differences with Seed-Coat Pattern and Pigmentation in *Phaseolus vulgaris, Genetics 8:* 552–560 (1923).
2. Hedges, B.R., R.G. Palmer, and L.A. Amberger, Electrophoretic Analyses of Soybean and Seed Proteins, in *Modern Methods of Plant Analysis New Series: Seed Analysis*, Vol. 14, edited by H.F. Linskens and J.F Jackson, Springer-Verlag, Berlin, 1992, pp. 143–158.
3. Botstein, D., R.L. White, M. Skolnick, and R.W. Davis, Construction of a Genetic Linkage Map in Man Using Restriction Fragment Length Polymorphisms, *Am. J. Hum. Genet. 32:* 314–331 (1980).
4. Mullis, K., F. Faloona, S. Scharf, R. Saiki, G. Horn, and H. Erlich, Specific Enzymatic Amplification of DNA *In Vitro*: The Polymerase Chain Reaction, *Cold Spring Harbor Symp. Quant. Biol. 51:* 263–273 (1986).
5. Litt, M., and J.A. Luty, A Hypervariable Microsatellite Revealed by *In Vitro* Amplification of a Dinucleotide Repeat Within the Cardiac Muscle Actin Gene, *Am. J. Hum. Genet. 44:* 397–401 (1989).
6. Tautz, D., Hypervariability of Simple Sequences as a General Source of Polymorphic DNA Markers, *Nucleic Acids Res. 17:* 6463–6471 (1989).
7. Weber, J.L., and P.E. May, Abundant Class of Human DNA Polymorphisms Which Can Be Typed Using the Polymerase Chain Reaction, *Am. J. Hum. Genet. 44:* 388–396 (1989).
8. Williams, J.K.G., A.R. Kubelik, K.J. Livak, J.A. Rafalski, and S.V. Tingey, DNA Polymorphisms Amplified by Arbitrary Primers Are Useful as Genetic Markers, *Nucleic Acids Res. 18:* 6531–6535 (1990).
9. Welsh, J., and M. McClelland, Fingerprinting Genomes using PCR with Arbitrary Primers, *Nucleic Acids Res. 18:* 7213–7218 (1990).
10. Vos, P., R. Hogers, M. Bleeker, M. Rijans, T. van der Lee, M. Hornes, A. Frijters, J. Pot, J. Peleman, M. Kuiper, and M. Zabeau, AFLP: A New Technique for DNA Fingerprinting, *Nucleic Acids Res. 23:* 4407–4414 (1995).
11. Keim, P., B.W. Diers, T.C. Olson, and R.C. Shoemaker, RFLP Mapping in Soybean: Association Between Marker Loci and Variation in Quantitative Traits, *Genetics 126:* 735–742 (1990).
12. Shoemaker, R.C., and T.C. Olson, Molecular Linkage Map of Soybean (*Glycine max* L. Merr.), in *Genetic Maps: Locus Maps of Complex Genomes*, edited by S.J. O'Brien, Cold Spring Harbor Laboratory Press, New York, 1993, pp. 6.131–6.138.
13. Shoemaker, R.C., and J.E. Specht, Integration of the Soybean Molecular and Classical Genetic Linkage Groups, *Crop Sci. 35:* 436–446 (1995).
14. Shoemaker, R., K. Polzin, J. Labate, J. Specht, E.C. Brummer, T. Olson, N. Young, V. Concibido, J. Wilcox, J.P. Tamulonis, G. Kochert, and H.R. Boerma, Genome Duplication in Soybean *(Glycine subgenus soja), Genetics 144:* 329–338 (1996).
15. Keim, P., J.M. Schupp, S.E. Travis, K. Clayton, T. Zhu, L. Shi, A. Ferreira, and D.M. Webb, A High-Density Soybean Genetic Map Based on AFLP Markers. *Crop Sci. 37:* 537–543 (1997).
16. Akkaya, M.S., A.A. Bhagwat, and P.B. Cregan, Length Polymorphism of Simple Sequence Repeat DNA in Soybean, *Genetics 132:* 1131–1139 (1992).
17. Morgante, M., and A.M. Olivieri, PCR-Amplified Microsatellites as Markers in Plant Genetics, *Plant J. 3:* 175–182 (1993).

18. Morgante, M., J.A. Rafalski, P. Biddle, S. Tingey, and A.M. Olivieri, Genetic Mapping and Variability of Seven Soybean Simple Sequence Repeat Loci, *Genome 37:* 763–769 (1994).
19. Rongwen, J., M.S. Akkaya, U. Lavi, and P.B. Cregan, The Use of Microsatellite DNA Markers for Soybean Genotype Identification, *Theor. Appl. Genet. 90:* 43–48 (1995).
20. Akkaya, M.S., R.C. Shoemaker, J.E. Specht, A.A. Bhagwat, and P.B. Cregan, Integration of Simple Sequence Repeat DNA Markers into a Soybean Linkage Map, *Crop Sci. 35:* 1439–1445 (1995).
21. Cregan P.B., T. Jarvik, A.L. Bush, R.C. Shoemaker, K.G. Lark, A.L. Kahler, N. Kaya, T.T. VanToai, D.G. Lohnes, J. Chung, and J.E. Specht, An Integrated Genetic Linkage Map of the Soybean Genome, *Crop Sci. 39:* 1464–1490 (1999).
22. Collins, F.S., L.D. Brooks, and A. Chakravarti, A DNA Polymorphism Discovery Resource for Research on Human Genetic Variation, *Genome Res. 8:* 1229–1231 (1998).
23. Kwok, P.-Y., Q. Deng, H. Zakeri, and D.A. Nickerson, Increasing the Information Content of STS-Based Genome Maps: Identifying Polymorphisms in Mapped STSs, *Genomics 31:* 123–126 (1996).
24. Lindblad-Toh, K., E. Winchester, M.J. Daly, D.G. Wang, J.N. Hirschhorn, J.P. Laviolette, K. Ardlie, D.E. Reich, E. Robinson, P. Sklar, N. Shah, D. Thomas, J.B. Fan, T. Gingeras, J. Warrington, N. Patil, T.J. Hudson, and E.S. Lander, Large-Scale Discovery and Genotyping of Single-Nucleotide Polymorphisms in the Mouse, *Nat. Genet. 24:* 381–385 (2000).
25. Cooper, D.N., B.A. Smith, H.J. Cooke, S. Niemann, and J. Schmidtke, An Estimate of Unique Sequence Heterozygosity in the Human Genome, *Hum. Genet. 69:* 201–205 (1985).
26. Li, W.H., and L.A. Sadler, Low Nucleotide Diversity in Man, *Genetics 129:* 513–523 (1991).
27. Harding, R.M., S.M. Fullerton, R.C. Griffiths, J. Bond, M.J. Cox, J.A. Schneider, D.S. Moulin, and J.B. Clegg, Archaic African and Asian Lineages in the Genetic Ancestry of Modern Humans, *Am. J. Hum. Genet. 60:* 772–789 (1997).
28. Wang, D.G., J.-B. Fan, C.-J. Siao, A. Berno, P. Young, R. Sapolsky, G. Ghandour, N. Perkins, E. Winchester, J. Spencer, L. Kruglyak, L. Stein, L. Hsie, T. Topaloglou, E. Hubbell, E. Robinson, M. Mittmann, M.S. Morris, N. Shen, D. Kilburn, J. Rioux, C. Nusbaum, S. Rozen, T.J. Hudson, R. Lipshutz, M. Chee, and E.S. Lander, Large-Scale Identification, Mapping and Genotyping of Single-Nucleotide Polymorphisms in the Human Genome, *Science 280:* 1077–1082 (1998).
29. Cho, R.J., M. Mindrinos, D.R. Richards, R.J. Sapolskym, M. Anderson, E. Drenkard, J. Dewdney, T.L. Reuber, M. Stammers, N. Federspiel, A. Theologis, W.H. Yang, E. Hubbell, M. Au, E.Y. Chung, D. Lashkari, B. Lemieux, C. Dean, R.J. Lipshutz, F.M. Ausubel, R.W. Davis, and P.J. Oefner, Genome-Wide Mapping with Biallelic Markers in *Arabidopsis thaliana, Nat. Genet. 23:* 203–207 (1999).
30. Bhattramakki, D., A. Ching, M. Morgante, M. Dolan, J. Register, H. Smith, S. Tingey, and J.A. Rafalski, Conserved Single Nucleotide Polymorphism (SNP) Haplotypes in Maize, *Abstracts of Plant-Animal Genome VIII* (2000).
31. Zhu, Y.L., S.M Hyatt, C.V. Quigley, Q.J. Song, N.D. Young, and P.B. Cregan, Single Nucleotide Polymorphism (SNPs) in Soybean Genes, ESTs, and Random Genomic Sequence, in *8th Biennial Conference on the Cellular and Molecular Biology of the Soybean*, 2000, p. PIII-08.

32. Risch, N., and K. Merikangas, The Future of Genetic Studies of Complex Human Diseases, *Science 273:* 1516 (1996).
33. Marshall, E., "Playing Chicken" over Gene Markers, *Science 278:* 2046–2048 (1997).
34. Marshall, E., Drug Firms to Create Public Database of Genetic Mutations, *Science 284:* 406–407 (1999).
35. Roberts, L., SNP Mappers Confront Reality and Find it Daunting, *Science 287:* 1898–1899 (2000).
36. Cregan P.B., A.A. Bhagwat, M.S. Akkaya, and Jiang Rongwen, Microsatellite Fingerprinting and Mapping of Soybean, *Methods Mol. Cell. Biol. 5:* 49–61 (1994).
37. Bassam B.J., G. Caetano-Anolles, and P.M. Gresshoff, Fast and Sensitive Staining of DNA in Polyacrylamide Gels, *Anal. Biochem. 196:* 80–83 (1991).
38. Diwan, N., and P.B. Cregan, Automated sizing of fluorescent labeled simple sequence repeat markers to assay genetic variation in soybean, *Thoer. Appl. Genet. 95:* 723–733 (1997).
39. Mian, M.A.R., T. Wang, D.V. Phillips, J. Alvernaz and H.R. Boerma, Molecular Mapping of the *Rcs3* Gene for Resistance to Frogeye Leaf Spot in Soybean, *Crop Sci. 39:* 1687–1691 (1999).
40. Song, Q., C.V. Quigley, T.E. Carter, R.L. Nelson, H.R. Boerma, J. Strachan, and P.B. Cregan, A Selected Set of Trinucleotide Simple Sequence Repeat Markers for Soybean Variety Identification, *Plant Var. Seeds 12:* 207–220 (1999).
41. Cheng, J., E.L. Sheldon, L. Wu, A. Uribe, L.O. Gerrue, J. Carrino, M. Heller, and J. O'Connell, Electric Field Controlled Preparation and Hybridization Analysis of DNA/RNA from *E. coli* on Microfabricated Bioelectronic Chips, *Nat. Biotechnol. 16:* 541–546 (1998).
42. Sosnowski, R., E. Tu, W. Butler, J. O'Connell, and M. Heller, Rapid Determination of Single Base Mismatch Mutations in DNA Hybrids by Direct Electric Field Control, *Proc. Natl. Acad. Sci. 94:* 1119–1123 (1997).
43. Braun, A., D.P. Little, B. Darnhofer-Demar, and H. Köster, Improved Analysis of Microsatellites Using Mass Spectrometry, *Genomics 46:* 18–23 (1997).
44. Kwok, P.-Y., and X. Chen, Detection of Single Nucleotide Variations, *Genet. Eng. 20:* 125–134 (1998).
45. Landegren, U., M. Nilsson, and P.-Y. Kwok, Reading Bits of Genetic Information: Methods for Single-Nucleotide Polymorphism Analysis, *Genome Res. 8:* 769–776 (1998).
46. Winzeler, E.A., D.R. Richards, A.R. Conway, A.L. Goldstein, S. Kalman, M.J. McCullough, J.H. McCusker, D.A. Stevens, L. Wodicka, D.J. Lockhart, and R.W. Davis, Direct Allelic Variation Scanning of the Yeast Genome, *Science 281:* 1194–1197 (1998).
47. Sapolsky, R., L. Hsie, A. Berno, G. Ghandour, M. Mittman, and J.B. Fan, High-Throughput Polymorphism Screening and Genotyping with High-Density Oligonucleotide Arrays, *Genet. Anal. Biomol. Eng. 14:* 187–192 (1999).
48. Michael, K.L., L.C. Taylor, S.L. Schultz, and D.R. Walt, Randomly Ordered Addressable High-Density Optical Sensor Arrays, *Anal. Chem. 70:* 1242–1248 (1998).
49. Ross, P., L. Hall, I. Smirnov, and L. Haff, High Level Multiplex Genotyping by MALDI-TOF Mass Spectrometry, *Nat. Biotechnol. 16:* 1347–1351 (1998).
50. Little, D.P., T.J. Cornish, M.J. O'Donnell, A. Braun, R.J. Cotter, and H. Köster, MALDI on a Chip: Analysis of Arrays of Low-Femtomole to Subfemtomole Quantities

of Synthetic Oligonucleotides and DNA Diagnostic Products Dispensed by a Piezoelectric Pipet, *Anal. Chem. 69:* 4540–4546 (1997).

51. Chen, J., M.A. Iannone, M.-S. Li, J.D. Taylor, P. Rivers, A.J. Nelson, K.A. Slentz-Kesler, A. Roses, and M.P. Weiner, A Microsphere-Based Assay for Multiplexed Single Nucleotide Polymorphism Analysis Using Single Base Extension, *Genome Res. 10:* 549–557 (2000).
52. Iannone, M.A., J.D. Taylor, J. Chen, M.-S. Li, P. Rivers, K.A. Slentz-Kesler, and M.P. Weiner, Multiplexed Single Nucleotide Polymorphism Genotyping by Oligonucleotide Ligation and Flow Cytometry, *Cytometry 39:* 131–140 (2000).
53. Tyagi S, and F.R. Kramer, Molecular Beacons: Probes That Fluoresce upon Hybridization, *Nat. Biotechnol. 14:* 303–308 (1996).
54. Tyagi S, D.P. Bratu, and F.R. Kramer, Multicolor Molecular Beacons for Allele Discrimination, *Nat. Biotechnol. 16:* 49–53 (1998).

Chapter 8

Using Natural Genes to Improve Soybean Quality

James R. Wilcox

USDA, Agricultural Research Service, Crop Production and Pathology Research, West Lafayette, IN 47907-1150

Introduction

Soybeans (*Glycine max*) have been grown in the United States, primarily for the oil and meal in the seed, since the mid-1940s. As demand for oil and meal increased, the area planted to soybeans has increased from 5.7 million hectares in 1944 to a projected 29 million hectares in 2000. Soybean oil has been recognized as a nutritious and economical vegetable oil, readily accepted by consumers in a variety of products. The meal is an exceptionally nutritious feed, widely accepted by the livestock industry. New markets are developing for soybean products in industry as well as in the food and feed industries. Opportunities are available to enhance the nutritional value of both the oil and meal through traditional breeding methods. Variability exists in germ plasm collections, or has been created by induced mutations to alter oil properties, increase seed protein, and reduce or eliminate antinutritional components of the meal. Incorporating this variability into improved soybean germ plasm will ensure that competitive soybean products exist in new and expanding world markets in the future.

Discussion

Oil Quality

Soybean oil is the major edible oil produced and consumed in the United States today. Because nearly all of the soybean oil produced goes into the edible oil market, emphasis on modifying oil properties should be directed toward improving the nutritional value, flavor, and stability of the oil. The nutritional value of soybean oil can be enhanced by reducing the saturated fatty acids that comprise ~150 g/kg of the oil. Palmitic acid (16:0), ~120 g/kg of the oil in commonly grown soybean cultivars, has been reduced to ~40 g/kg by combining alleles at different loci that control synthesis of this fatty acid (Table 1) (1–4). The low palmitate trait has been incorporated into soybean cultivars, and a low saturated soybean oil, with half the saturated fat of most vegetable oil, is currently available to consumers.

Genetic variability is also available to increase the saturated fatty acids in soybean oil (Table 1). Germ plasm has been developed with as much as 400 g/kg

TABLE 1
Range in Fatty Acid Contents of Soybean Oil Due to Combinations of Naturally Occurring or Mutant Alleles

Fatty acid	Low (g/kg)	Reference	High (g/kg)	Reference
Palmitic	40	1	280	30
		31		
		4		
Stearic	32	32	304	33
Oleic	163	34	584	PC[a]
Linoleic	267	PC	640	31
Linolenic	25	31	159	35

[a]PC, J.W. Burton, personal communication.

palmitic acid and 300 g/kg stearic acid in the oil. Soybean oil with high saturates may be particularly valuable in shortenings or in specialty oil markets.

The oil of commonly grown soybean cultivars typically contains 16–24 g/kg oleic acid. Long-term breeding programs by USDA-ARS scientists at Raleigh, NC, have been successful in developing germ plasm with nearly 600 g/kg oleic acid in the oil (J.W. Burton, personal communication). There is invariably a strong inverse, and proportional, relationship between oleic and linoleic acid in soybean oil. The high oleic soybean oil typically contains ~260 g/kg linoleic acid (Table 1).

The impetus for much of the research on the genetics of oil quality was prompted by interest in reducing linolenic acid in soybean oil. Reductions in this fatty acid would improve oil stability and flavor, reducing the need for hydrogenation of the oil. Mutation breeding was successful in creating soybean germ plasm with ~35 g/kg linolenic acid, half the amount in the oil of commonly grown cultivars. Further breeding efforts, involving both mutation and selection among progeny of crosses between low 18:3 germ plasm lines, have reduced linolenic acid to <25 g/kg (Table 1).

Most of the genetic changes in oil composition are due to alleles with additive effects; thus, combining specific alleles can result in major changes in the fatty acid composition of the oil. The alleles are simply inherited and can be readily incorporated into new cultivars, with the potential to produce "designer" oils with specific characteristics.

Protein Accumulation

Protein comprises ~400 g/kg of the soybean seed on a moisture-free basis. Soy meal, with its high protein concentration, is commonly the economically more valuable portion of the seed compared with soybean oil. Soy meal is a major component of livestock feeds, and is used extensively in poultry rations.

Genetic variability for seed protein concentration ranges as high as 500–520 g/kg on a moisture-free basis. Increasing seed protein should enhance the nutritional value of soy meal, and this genetic modification of soybean can be readily accomplished. However, there are two major deterrents to developing soybean with high seed protein.

First, there is a moderately strong inverse relationship between seed protein concentration and seed yield. It has been very difficult to develop soybean cultivars with seed protein concentration in excess of 470 g/kg and with economically profitable seed yield. Second, there is a strong inverse relationship between seed oil and seed protein concentration. As seed protein increases, oil, the other valuable component of the seed, decreases, further reducing the profitability of high protein soybean.

The inverse relationship between seed protein and seed yield can be overcome with intensive breeding efforts. A seed protein concentration of 472 g/kg was transferred, by backcrossing, to a soybean cultivar that originally contained 408 g/kg protein and 204 g/kg seed oil (5). The yield of the cultivar was fully recovered, and the backcross-derived line averaged 174 g/kg seed oil (Table 2). The economic value of the high-protein, backcross-derived line, based on seed yield and the value of the meal and oil, was greater than that of the original cultivar. This research clearly demonstrated that economically profitable soybean cultivars with high seed protein concentration can be developed.

Soybean breeders have intensified their efforts to increase seed protein while maintaining profitable seed yields. Excellent progress has been made, and typically soybean lines with 470 g/kg or greater seed protein have produced seed yields equal to 90% of the yields of currently grown cultivars. Recently, breeding lines have been developed with increased seed protein and with yields comparable to those of currently grown, publicly developed soybean cultivars (Table 3).

Seed Coat Peroxidase

Soybean seeds differ in activity of the enzyme peroxidase that occurs in the seed coat (6). Cultivars with "high" activity have 100-fold greater specific activity than cultivars with "low" peroxidase activity (7). Differences in peroxidase activity in soybean seed coats are due to alleles at a single locus, and the high activity trait can be readily incorporated into improved soybean cultivars (8).

For many years, differences in peroxidase activity were simply used as a trait to aid in cultivar identification. Soybean peroxidase has high thermal stability (>85°C), and

TABLE 2
Performance of Successive Backcross Progenies When Transferring High Seed Protein to a High-Yielding Soybean Cultivar

Strain	Cultivar genotype (%)	Yield (kg/ha)	Protein (g/kg)	Oil (g/kg)
Donor	0	800	498	148
Backcross 1	50	2145	474	175
Backcross 2	75	2179	476	167
Backcross 3	88	2529	466	171
Backcross 4	94	2831	472	174
Cultivar	100	2461	408	204

TABLE 3
Performance of High-Protein Breeding Lines Compared with Currently Grown Soybean Cultivars

Strain	Yield (kg/ha)	Mature (date)	Protein (g/kg)	Oil (g/kg)
Maturity Group I				
CX1654-31	2410	8-27	479	162
Marcus 95	2350	8-27	416	202
Maturity Group II				
CX1654-27	2640	9-10	474	160
CX1654-92	2580	9-10	462	162
IA2038	2660	9-12	414	196

new uses for soybean peroxidase have recently been identified. These include formaldehyde-free phenolic resins, medical diagnostics, engineering of plastics, and dough conditioning (9). Five tons of soybean hulls are being exported to Israel every month so that peroxidase can be extracted from the hulls for use in the baking industry. These new uses enhance the economic value of what was once considered a by-product of soybean. Currently, all advanced breeding lines in public breeding programs are evaluated to determine whether they contain high or low peroxidase activity in the seed coat.

Trypsin Inhibitors

Trypsin inhibitors in soybean cause pancreatic hypertrophy and growth inhibition. The primary protease inhibitors are the Kunitz and the Bowman-Birk, and heat treatment of soybean meal is necessary to inactivate these inhibitors. The Kunitz inhibitor is controlled by alleles at a single locus and the recessive allele, *ti*, results in the absence of this inhibitor (10). A commercially available cultivar "Kunitz" has been developed that lacks this trypsin inhibitor (11).

Germ plasm lacking the Bowman-Birk trypsin inhibitor has been identified in accessions of several perennial soybean species (12) (Table 4). *Glycine tomentalla*, one of the perennial species lacking the Bowman-Birk inhibitor, has been successfully crossed with the cultivated *G. max*. Transferring null alleles from *G. tomentalla* for the Bowman-Birk trypsin inhibitor to *G. max* creates the opportunity to develop soybeans lacking both of the major trypsin inhibitors. This would reduce or eliminate the need for heat treatment to inactivate these growth inhibitors when soybean is used as a livestock feed.

Lipoxygenases

Soybean contains three lipoxygenases that are involved in oxidation of several substrates and contribute to off-flavors of various soy products. Heat treatment is used to suppress lipoxygenase activity, but this can cause problems in solubility of soy proteins, and both odor and flavor problems. Germ plasm that lacks each of the three

TABLE 4
Cultivated and Perennial *Glycine* Species Lacking Trypsin Inhibitors[a]

Species	Trypsin inhibitor	Reference
Glycine max cv "Kunitz"	Kunitz	10
Glycine curvata	Bowman-Birk	12
G. cyrtoloba	"	"
G. latifolia	"	"
G. microphylla	"	"
G. tabacina	"	"
G. tomentalla	"	"

[a]*Source:* Ref. 12.

lipoxygenases has been identified, and the null-alleles are simply inherited (13–16) (Table 5). Germ plasm was developed that lacked two of the lipoxygenase enzymes, –L1–L3 and –L2–L3. Because of a close repulsion linkage between genetic loci controlling L1 and L2, genetic combinations lacking L1 and L2 or the triple nulls were not identified until the early 1990s (17). At present, cultivars are available that lack all three of these lipoxygenases. Data have demonstrated that L2 and L3 are the primary causes of flavor problems associated with these lipoxygenase enzymes (18,19).

Phytic Acid

Phytic acid (myo-inositol 1,2,3,4,5,6-hexaphosphate) is the primary storage form of phosphate in soybean seed, and phytic acid phosphorous is nutritionally unavailable to nonruminant livestock (20). The unavailable phosphorous in soy meal is excreted by livestock and has become a serious pollutant in intensive livestock feeding operations. In addition, phytic acid is a strong chelating agent that can bind metal ions, reducing the availability of zinc, calcium, iron, and magnesium (21,22). Soy meal is commonly supplemented with inorganic phosphorous or with phytase to increase the bioavailability of phosphorous and of these metal ions (23–25).

Recent research has identified soybean germ plasm with reduced levels of phytic acid and with increased inorganic phosphorous in the seed (26). The low-phytate germ

TABLE 5
Alleles Conditioning the Presence or Absence of Lipoxygenases in Soybeans

Lipoxygenase	Genotype	Reference
–L1	*l×1 L×2 L×3*	15
–L2	*L×1 l×2 L×3*	13
–L3	*L×1 L×2 l×3*	16
–L1, +L2, +L3	*l×1 L×2 L×3*	14
–L1, +L2, –L3	*l×1 L×2 l×3*	14
+L1, –L2, –L3	*L×1 l×2 l×3*	14
–L1, –L2, –L3	*l×1 l×2 l×3*	17

plasm was created using a chemical mutagen, and breeding lines with inherently high inorganic phosphorous and low phytic acid were selected from among thousands of mutant progenies. The cultivar Athow and a nonmutated breeding line averaged 23.81 and 26.82 g/kg phytic acid, respectively, in the seed (Table 6). Only 15 and 11% of the seed phosphorous in these two lines, respectively, was inorganic P, and would be nutritionally available to livestock. The mutant line M153 has ~66% inorganic P, and a second line, M766, ~48% inorganic P, or "available P" in the seed.

The low-phytate soybean complements research that was successful in reducing phytic acid and increasing inorganic phosphorous in seed of grain crops including corn, barley, and rice (27–29). Swine and poultry feed rations commonly contain grain and soy components. Use of low-phytate soybean in feed rations, combined with low-phytate corn or small grains, would represent an improved source of phosphorous for livestock feeds, reducing the need for phosphorous or phytase supplementation. Additionally, the use of low-phytate feeds would be a valuable tool to reduce the amount of phosphorous in livestock wastes, reducing the adverse effects of livestock feeding operations on the environment.

Summary

The cultivated soybean contains a wealth of naturally occurring variability that can be used to improve compositional traits of this crop. This genetic variability has been used primarily to increase protein accumulation in soybean. Closely related perennial species of *Glycine* expand the variability available to plant breeders to include traits such as the absence of trypsin inhibitors. Perennial soybean species are a crucial resource for unique traits, and expanded research efforts are needed to capitalize on this genetic resource. When needed variability for economically important traits is not readily available, soybean breeders have created mutations for these traits within the soybean genome. Modifications in fatty acid composition of soy oil and creation of low-phytate soybean germ plasm are examples of successful mutation breeding. These genetic changes within the soybean genome are readily accepted by the soybean industry and by consumers and can ensure the continued profitability of soybean products.

TABLE 6
Seed Phytic Acid, Phytic Acid P and Inorganic P of the Cultivar Athow and of Selected Soybean Mutants

Cultivar or mutant line	Phytic acid (g/kg)	Phytic acid P (g/kg)	Inorganic P (g/kg)	Phytic acid P + Inorganic P (g/kg)	Inorganic P (%)
Athow	23.81	4.32	0.74	5.06	15
Nonmutant	26.82	4.86	0.61	5.47	11
Mutant M153	10.47	1.90	3.62	5.52	66
Mutant M766	16.76	3.07	2.79	5.86	48

References

1. Fehr, W.R., G.A. Welke, E.G. Hammond, D.N. Duvick, and S.R. Cianzio, Inheritance of Reduced Palmitic Acid Content in Seed Oil of Soybean, *Crop Sci. 31:* 88–89 (1991).
2. Horejsi, T.F., W.R. Fehr, G.A. Welke, D.N. Duvick, E.G. Hammond, and S.R. Cianzio, Genetic Control of Reduced Palmitate Content in Soybean, *Crop Sci. 34:* 331–334 (1994).
3. Stojsin, D., G.R. Ablett, B.M. Luzzi, and J.W. Tanner, Use of Gene Substitution Values to Quantify Partial Dominance in Low Palmitic Acid Soybean, *Crop Sci. 38:* 1437–1441 (1998).
4. Wilcox, J.R., J.W. Burton, G.J. Rebetzke, and R.F. Wilson, Transgressive Segregation for Palmitic Acid in Seed Oil of Soybean, *Crop Sci. 34:* 1248–1250 (1994).
5. Wilcox, J.R. and J.F. Cavins, Backcrossing High Seed Protein to a Soybean Cultivar, *Crop Sci. 35:* 1036–1041 (1995).
6. Buttery, B.R., and R.I. Buzzell, Peroxidase Activity in Seeds of Soybean Varieties, *Crop Sci. 8:* 722–725 (1968).
7. Gijzen, M., R. van Huystee, and R.I. Buzzell, Soybean Seed Coat Peroxidase, *Plant Physiol. 103:* 1061–1066 (1993).
8. Buzzell, R.I., and B.R. Buttery, Inheritance of Peroxidase Activity in Soybean Seed Coat, *Crop Sci. 9:* 387–388 (1969).
9. Vierling, R.A., Soybean, New Uses, in *Proceedings of the 28th Soybean Seed Research Conference,* American Seed Trade Association, Washington, D.C., pp. 64–72 (1998).
10. Orf, J.H., and T. Hymowitz, Inheritance of the Absence of the Kunitz Trypsin Inhibitor in Seed Protein of Soybeans, *Crop Sci. 19:* 107–109 (1979).
11. Bernard, R.L., T. Hymowitz, and C.R. Cremeens, Registration of 'Kunitz' Soybean, *Crop Sci. 31:* 232–233 (1991).
12. Domagalski, J.M., K.P. Kollipara, A.H. Bates, D.L. Brandom, M. Friedman, and T. Hymowitz, Nulls for the Major Soybean Bowman-Birk Protease Inhibitor in the Genus Glycine, *Crop Sci. 32:* 1502–1505 (1992).
13. Davies, C.S., and N.C. Nielsen, Genetic Analysis of a Null-Allele for Lipoxygenase-2 in Soybean, *Crop Sci. 26:* 460–463 (1986).
14. Davies, C.S., and N.C. Nielsen, Registration of Soybean Germplasm That Lacks Lipoxygenase Isozymes, *Crop Sci. 87:* 370–371 (1987).
15. Hildebrand, D.F., and T. Hymowitz, Inheritance of Lipoxygenase-1 Activity in Soybean Seeds, *Crop Sci. 22:* 851–853 (1982).
16. Kitamura, K., C.S. Davies, N. Kaizuma, and N.C. Nielsen, Genetic Analysis of a Null-Allele for Lipoxygenase-3 in Soybean Seeds, *Crop Sci. 23:* 924–927 (1983).
17. Kitamura, K., Breeding Trials for Improving the Food-Processing Quality of Soybeans, *Trends Food Sci. Technol. 4:* 64–67 (1993).
18. Davies, C.S., S.S. Nielsen, and N.C. Nielsen, Flavor Improvement of Soybean Preparations by Genetic Removal of Lipoxygenase, *J. Am. Oil Chem. Soc. 64:* 1428–1433 (1987).
19. Kobayashi, A., Y. Tsuda, N. Hirata, K. Kubota, and K. Kitamura, Aroma Constituents of Soybean [*Glycine max* (L.) Merrill] Milk Lacking Lipoxygenase Isozymes, *J. Agric. Food Chem. 43:* 2449–2452 (1995).
20. Erdman, J.W., Jr., Oilseed Phytates: Nutritional Implications, *J. Am. Oil Chem. Soc. 56:* 736–741 (1979).

21. Erdman, J.W., Bioavailability of Trace Minerals from Cereals and Legumes, *Cereal Chem. 58:* 21–26 (1981).
22. McChance, R.A., and E.M. Widdowson, Phytic Acid in Human Nutrition, *Biochem. J. 29:* 42695–42699 (1935).
23. Adeola, A., Digestive Utilization of Minerals by Weanling Pigs Fed Copper- and Phytase-Supplemented Diets, *Can. J. Anim. Sci. 75:* 603–610 (1995).
24. Adeola, A., B.V. Lawrence, A.L. Sutton, and T.R. Cline, Phytase-Induced Changes in Mineral Utilization in Zinc-Supplemented Diets for Pigs, *J. Anim. Sci. 73:* 3384–3391 (1995).
25. Mendoza, C., F.E. Viteri, B. Lonnerdal, K.A. Young, V. Baroy, and K.H. Brown, Effect of Genetically Modified, Low-Phytic Acid Maize on Absorption of Iron from Tortillas, *Am. J. Clin. Nutr. 68:* 1123–1127 (1998).
26. Wilcox, J.R., G.S. Premachandra, K.A. Young, and V. Raboy, Isolation of High Seed Inorganic P, Low-Phytate Soybean Mutants, *Crop Sci. 40:* 1601–1605 (2000).
27. Ertl, D.S., K.A. Young, and V. Raboy, Plant Genetic Approaches to P Management in Agricultural Production, *J. Environ. Qual. 27:* 299–304 (1998).
28. Larson, S.R., K.A. Young, A. Cook, T.K. Blake, and V. Raboy, Linkage Mapping of Two Mutations That Reduce Phytic Acid Content of Barley Grain, *Theor. Appl. Genet. 97:* 141–146 (1998).
29. Raboy, V., and P. Gerbasi, Genetics of Myo-inositol Phosphate Synthesis and Accumulation, in *Myo-inositol Phosphates, Phosphoinositides, and Signal Transduction,* edited by B.B. Biswas and S. Biswas, Plenum Press, New York, 1996, pp. 257–285.
30. Fehr, W.R., G.A. Welke, E.G. Hammond, D.N. Duvick, and S.R. Cianzio, Inheritance of Elevated Palmitic Acid Content in Soybean Seed Oil, *Crop Sci. 31:* 1522–1524 (1991).
31. Stojsin, D., B.M. Luzzi, G.R. Ablett, and J.W. Tanner, Inheritance of Low Linolenic Acid Level in the Soybean Line RG10, *Crop Sci. 38:* 1441–1444 (1998).
32. Graef, G.L., W.R. Fehr, and E.G. Hammond, Inheritance of Three Stearic Acid Mutants of Soybean, *Crop Sci. 25:* 1076–1079 (1985).
33. Hammond, E.G., and W.R. Fehr, Registration of A6 Germplasm Line of Soybean, *Crop Sci. 23:* 192–193 (1983).
34. Nickell, A.D., J.R. Wilcox, and J.F. Cavins, Genetic Relationships Between Loci Controlling Palmitic and Linolenic Acids in Soybean, *Crop Sci. 31:* 1169–1171 (1991).
35. Takagi, Y., A.B.M. Mamun Hossain, T. Yanagita, and S. Kusaba, High Linolenic Acid Mutant in Soybean Induced by X-Ray Irradiation, *Jpn. J. Breed. 39:* 403–409 (1989).

Chapter 9

Enhancing Disease Resistance of Crops Through Breeding and Genetics

J.B. Holland

USDA-ARS Plant Science Research Unit, Department of Crop Science, North Carolina State University, Raleigh, NC 27695-7620

Introduction

Flor (1,2) was the first to recognize that the genetic basis of plant disease often depends on the genes carried by the pathogen as well as those in the host plant. The gene-for-gene hypothesis describes the complementarity of resistance genes (R-genes) in the host plant with avirulence genes in the pathogen. "For each gene that conditions resistance in the host there is a corresponding gene that conditions pathogenicity in the parasite" (3). A plant's R-gene is effective only against those isolates or races of the pathogen that possess the complementary avirulence gene (Tables 1 and 2). Flor (2) postulated that the products of host R-genes and pathogen avirulence genes interact physically to trigger the resistance response. Without this physical interaction, the resistance response does not occur, and the pathogen is able to invade the tissue of the host plant and cause disease. Thus, resistance is an active response on the part of the plant to infection by a specific pathogen. On the other hand, disease can result when no such active response occurs, as when the host lacks the appropriate R-gene, or the pathogen lacks any avirulence genes whose products might be detected by the host (Tables 1 and 2). Pathogen races are defined by which avirulence genes they carry; for this reason, gene-for-gene resistances are termed "race specific."

This complementarity of host and pathogen genes has profound implications for the development of long-lasting resistance and disease management. It follows

TABLE 1
Hypothetical Outcomes of Interactions Between Pathogens with One Avirulence Gene (*A*) and Host Plants with One Resistance Gene (*R*) in a Gene-for-Gene Interaction System

	Host genotype	
Pathogen genotype	*RR* or *Rr*	*rr*
AA or *Aa*	Resistance response, no disease	Susceptibility, disease
aa	Susceptibility, disease	Susceptibility, disease

TABLE 2
Hypothetical Outcomes of Interactions Between Pathogens with Three Avirulence Genes (A_1, A_2, A_3) and Host Plants with Three Resistance Genes (R_1, R_2, R_3) in a Gene-for-Gene Interaction System

	Host genotype[a]							
Pathogen genotype[a]	$R_{1\_}R_{2\_}R_{3\_}$	$R_{1\_}R_{2\_}r_3r_3$	$R_{1\_}r_2r_2R_{3\_}$	$r_1r_1R_{2\_}R_{3\_}$	$R_{1\_}r_2r_2r_3r_3$	$r_1r_1R_{2\_}r_3r_3$	$r_1r_1r_2r_2R_{3\_}$	$r_1r_1r_2r_2r_3r_3$
$A_{1\_}A_{2\_}A_{3\_}$	Resistance response, no disease	Resistance response, no disease	Resistance response, no disease	Resistance response, no disease	Resistance response, no disease	Resistance response, no disease	Resistance response, no disease	Susceptibility, disease
$A_{1\_}A_{2\_}a_3a_3$	Resistance response, no disease	Resistance response, no disease	Resistance response, no disease	Resistance response, no disease	Resistance response, no disease	Resistance response, no disease	Susceptibility, disease	Susceptibility, disease
$A_{1\_}a_2a_2A_{3\_}$	Resistance response, no disease	Resistance response, no disease	Resistance response, no disease	Resistance response, no disease	Resistance response, no disease	Susceptibility, disease	Resistance response, no disease	Susceptibility, disease
$a_1a_1A_{2\_}A_{3\_}$	Resistance response, no disease	Resistance response, no disease	Resistance response, no disease	Resistance response, no disease	Susceptibility, disease	Resistance response, no disease	Resistance response, no disease	Susceptibility, disease
$A_{1\_}a_2a_2a_3a_3$	Resistance response, no disease	Resistance response, no disease	Resistance response, no disease	Susceptibility, disease	Resistance response, no disease	Susceptibility, disease	Susceptibility, disease	Susceptibility, disease
$a_1a_1A_{2\_}a_3a_3$	Resistance response, no disease	Resistance response, no disease	Susceptibility, disease	Resistance response, no disease	Susceptibility, disease	Resistance response, no disease	Susceptibility, disease	Susceptibility, disease
$a_1a_1a_2a_2A_{3\_}$	Resistance response, no disease	Susceptibility, no disease	Resistance response, no disease	Resistance response, no disease	Susceptibility, disease	Susceptibility, disease	Resistance response, no disease	Susceptibility, disease
$a_1a_1a_2a_2a_3a_3$	Susceptibility, disease	Susceptibility, disease	Susceptibility, disease	Susceptibility, disease	Susceptibility, disease	Susceptibility, disease	Susceptibility, disease	Susceptibility, disease

[a] $R_{x\_}$ includes both R_xR_x and R_xr_x genotypes; $A_{x\_}$ includes both A_xA_x and A_xa_x genotypes.

from the gene-for-gene hypothesis that host plant R-genes will be effective only insofar as they correspond to avirulence genes that occur in high frequency in the pathogen populations. In theory, at least, humans can control the deployment of R-genes across vast areas; but humans cannot easily dictate the genetic make-up of pathogen populations. If pathogen populations are heterogeneous at avirulence loci, then deployment of gene-for-gene type resistances in cultivars that are grown widely will exert strong selection pressure for loss of avirulence to the deployed genes in the pathogen population (4). Pathogen populations respond to such selection pressure with a decrease in frequency of the complementary avirulence gene. The result often is that after a few years of use, gene-for-gene types of resistance lose effectiveness, or "break down" (5). In fact, nothing "happens" to the R-gene, but changes in the pathogen population reduce the frequency of avirulence products detectable by the R-gene; thus, the resistance response is not triggered and disease results.

Despite their general lack of durability, gene-for-gene types of resistance are frequently exploited by plant breeders for one primary reason, i.e., they are easy to identify and to incorporate into elite cultivars. With good inoculation methods, it is easy to identify and eliminate the susceptible plants in a breeding program, and conversely to identify the resistant plants, which are maintained for continued backcrossing into an elite cultivar, or for line development. Many gene-for-gene types of resistance are available in many crops, and as long as the supply exists, they will be used. The situation is something like the excessive use of fossil-fuels in many societies; we know they will not last forever, but as long as they exist, they are regarded as the cheapest and easiest source to use. In oat (*Avena sativa*), major genes for resistance to the crown rust pathogen, *Puccinia coronata*, were deployed extensively and quickly lost. By the 1960s (after only about half a century of their use), all major R-genes in the cultivated species were rendered ineffective to the crown rust populations of the North Central region of the United States (6). However, because the genes are so simple to work with, new R-genes identified in a wild relative, *A. sterilis*, were introgressed into cultivated types that were then widely released. The wild species–derived major gene resistances were no more durable than those from the cultivated species, but they continue to be used to provide at least a few years of resistance to this destructive disease (7,8). Similar "boom-and-bust" cycles were encountered with potato (*Solanum tuberosum*) blight (*Phytopthera infestans*), coffee (*Coffea arabica*) rust (*Hemileia vastatrix*) (4), powdery mildew (*Erysiphe graminis*) of wheat (*Triticum aestivum*), and barley (*Hordeum vulgare*) (9).

The gene-for-gene situation described does not apply to all plant-pathogen interactions, however. For some pathogens, no variability for virulence is known, and against these types of pathogens, single genes conferring complete resistance have maintained their effectiveness over relatively long periods of time (10). For other pathogens, there are a limited number of races, and gene-for-gene type resistances may eventually break down, but they have considerably longer life spans.

Typically, pathogens that are obligate biotrophs and highly specialized such that they can survive and reproduce only on the crop that they attack (or are restricted to a few hosts), tend to form races with different virulence capabilities more frequently than pathogens that have broader host ranges (10). Examples of highly specialized biotrophs include the fungi that cause rust diseases of cereals and flax (*Puccinia* species and *Melampsori lini*, respectively), mildews of lettuce and cereals (*Bremia lactucae* and *Erysiphe graminis*, respectively), and potato blight (*Phytopthera infestans*). More generalized pathogens, such as viruses and vascular wilt fungi (*Fusarium oxysporum*) form races much less frequently (10). For those pathogens that do not easily overcome gene-for-gene type resistances, such resistance will continue to be used effectively and will be manipulated relatively easily by traditional plant breeding procedures. It is the specialized, obligate biotrophic pathogens (particularly those that regularly reproduce sexually) that present the greatest present and future threat to crop production, and against which the gene-for-gene types of resistance tend to be quickly nullified. What prospects exist for the durable control of crop diseases caused by such pathogens? Traditional plant breeding offers two potential solutions, i.e., pyramiding R-genes and exploiting partial resistance. Molecular genetics techniques have offered some new tools for enhancing disease resistance, including DNA markers and marker-assisted selection, gene isolation and analysis techniques, and genetic transformation. Technologies that are just appearing on the horizon, such as chimeraplasty or direct gene modification, may eventually prove very powerful for developing durable resistance that avoids some of the controversy attending to genetically modified crops. Finally, crop management and rotations, and attention to the agroecological milieu should remain important aspects of disease management. Each of these alternatives or possibilities are discussed in turn.

Discussion

Pyramiding Disease Resistance Genes

One way to increase the durability of race-specific R-genes is to combine multiple genes with resistances to different spectra of the pathogen population into a single genotype. Deployment of R-gene combinations is predicted to be more durable than individual R-genes because to overcome resistance, pathogens must acquire mutations at multiple avirulence loci (10). Backcrossing types of schemes can also be used to introgress multiple R-genes into a single cultivar, but screening for resistance to more than one race and selection for multiple genes add complexity to the process. Furthermore, the useful life span of R-genes is not necessarily greater when they are deployed in combinations than when they are deployed individually. Although some R-gene combinations (e.g., stem rust resistance in U.S. winter wheat) have remained effective over many years (8,11), others have been quickly rendered ineffective, e.g., crown rust resistance in oat (12), and powdery mildew and yellow rust (*P. striiformis*) resistances in European wheat (9,13). Pathogens that have a sexual life stage tend to be able to overcome gene combinations more quickly because of the recombination that occurs in sexual reproduction (8).

R-genes can also be deployed in combinations in different genetic backgrounds. This can simplify some of the technical difficulties of gene pyramiding in a single cultivar but may introduce other problems. Cultivar mixtures including cultivars with different R-genes offer broad resistance against pathogen populations (14,15). Cultivar mixtures are sometimes more heterogenous than farmers or grain processors would like, however. For example, a mixture of cultivars that mature at substantially different times may be difficult to harvest mechanically, or the end-use characteristics of a mixture may be too variable to permit marketing for specific processing purposes. Nevertheless, mixtures with adequate uniformity can sometimes be developed without introducing complexities into the breeding process (16).

A multiline variety is a group of lines that are highly related but differ specifically for which R-genes they carry (17). The background genetic uniformity of the lines provides agronomic uniformity, whereas the variation at resistance loci provides protection against diverse pathogen populations. The major drawback of multiline varieties is the great amount of effort required to develop them (18).

Partial Resistance

Partial resistance is resistance that does not fully exclude the pathogen from reproducing on the host, but reduces the amount of disease damage that is done by the pathogen. Partial resistance can also slow the progress of disease epidemics. If disease damage is limited, economic losses in yield and quality may be minimized. The primary advantage of partial resistance is that it should be more durable than race-specific, complete resistance, i.e., selection pressure on the pathogen population is reduced because the resistance does not completely exclude specific genotypes from reproducing (19). Partial resistance may, nevertheless, be race specific (10,20). There is some evidence that pathogen populations can adapt to partial resistance, but the evolution of virulence to partial resistance is expected to be much slower than to complete resistance (10,21). Durability of resistance is impossible to prove except in retrospect, but Stuthman (22) noted that the oat cultivar "Portage" has maintained its high level of partial resistance to crown rust from the time of its release in 1960 to the present.

Two major impediments to selection for partial resistance to rust diseases are the difficulty in measuring partial resistance accurately and the often quantitative inheritance of partial resistance. Sometimes partial resistance can be assessed by a simple measurement of terminal disease severity in field plots, but in other cases, it is difficult to distinguish different levels of partial resistance with a single rating. In the latter case, area under the disease progress curve (AUDPC), based on measurements of the percentage of leaf area infected that are made periodically during the growing season, is a useful measure of partial disease resistance in the field, but is very labor intensive to measure. Partial resistance can also be characterized by its components, including latent period, infection efficiency, infectious period, and spore production (23,24), but this is also labor intensive. Partial resistance is usually polygenically inherited and often has low heritability, although there are exceptions to this (19,23,24). Well-characterized exceptions to polygenically inherited partial resistance include the Lr34 gene, which confers adult plant resistance to wheat leaf rust (25). The amount of field screening and labor required to measure partial resistance accurately have hindered the more general exploitation of partial resistance. One alternative to exhaustive phenotypic evaluation of partial resistance is selection in populations lacking gene-for-gene resistance for genotypes that yield well under both inoculated and disease-free conditions (created, for example, by treatment with systemic pesticides). This can result in simultaneous improvement in yield potential and partial resistance with less labor (Holland, J.B., and G.P. Munkvold, unpublished data).

DNA Markers: Mapping Resistance Genes

The development of DNA markers that allow unambiguous determination of which allele a plant or line has inherited at a specific chromosomal location has greatly increased our understanding of the genomic organization of R-genes and may be useful in breeding for durable resistance. By tracking simultaneously the inheritance of alleles at DNA marker loci and disease resistance or susceptibility phenotypes in segregating populations, one can determine which DNA markers are physically linked on the chromosome to the gene(s) conferring resistance (see, e.g., 26,27). If a marker allele is always co-inherited with the resistance phenotype, then the marker gene must be linked tightly to the R-gene. On the other hand, if there is no more than a random association between the marker allele and the resistance phenotype, the marker must be on a different chromosome or distantly located on the same chromosome as the R-gene. The level of association between marker genes and R-genes can be intermediate between these extremes and is quantified in terms of the frequency of recombination that is observed between the two genes. Recombination is a natural result of meiosis, and recombination frequency can range from 0 (tightly linked) to 50% (unlinked). Marker genes that are more tightly linked to the R-gene are more useful for breeding because selection on the basis of marker genotypes can be performed with higher probability that the selected plants or lines also have the desired resistance allele. DNA marker–assisted selection can be more effective than phenotypic selection if the disease resistance phenotype is difficult or costly to evaluate; it can reduce the number of generations required to backcross a R-gene into an elite cultivar background; it can simplify selection for R-gene combinations, thus making gene pyramiding a more viable option in cultivar development (28,29). Nevertheless, there are cases in which disease resistance can be easily and reliably scored phenotypically; in these situations, DNA markers may be a more expensive and less reliable breeding tool!

DNA markers can also be used to identify the chromosomal locations of genes conferring partial resistance (30). In this case, because the resistance phenotype is quantitative rather than discrete and qualitative, it is not easy to estimate accurately the recombination frequency between the marker and resistance loci. Nevertheless, associations between DNA marker genes and partial resistance genes can be tested by estimating the average quantitative resistance value of all of the plants or lines that have the same genotypes at the marker locus, and by comparing the means of the different marker genotypic classes. If there is no significant difference among the classes, then the marker gene is not linked to genes conferring partial resistance, or if it is, the effect is so small as to remain undetected. On the other hand, if one marker genotype group has significantly greater partial resistance than the other genotypic group at the same locus, the marker gene is considered to be linked to a quantitative trait locus (QTL), affecting partial resistance. The precise position of QTL relative to DNA markers is not easy to resolve, but recombination frequencies between marker genes and most likely QTL positions can be determined, and the most closely linked DNA markers can be used for indirect selection of partial resistance (29,30). Extensive phenotypic evaluation is required to characterize the different levels of partial resistance in the mapping population, but once this is done and the marker-resistance associations have been established, marker-assisted selection can be used to incorporate the partial resistance alleles at multiple loci without requiring phenotypic evaluations during each new generation (although the final products of such breeding efforts would surely require field evaluation to confirm their resistance phenotype and agronomic utility). Furthermore, marker-assisted selection can be used to pyramid both partial resistance alleles and gene-for-gene resistance alleles into a single genotype, which can be extremely difficult using phenotypic selection (31). Combining multiple gene-for-gene and partial resistances could provide complete resistance for some time, followed by partial resistance if the gene-for-gene resistances break down.

DNA Markers: Genomic Organization of Resistance Genes

In addition to their use in applied cultivar development programs, DNA markers have also proven useful for understanding the genomic organization of R-genes. Classical genetic studies in maize (*Zea mays*) of resistance to common rust (*Puccinia sorghi*) revealed that many R-genes mapped to a small region on chromosome 10 (32). Thus, the R-genes form a cluster in this region, rather than being distributed randomly throughout the genome. It was hypothesized that the genes conferring resistance to different isolates within this cluster had related functions, and likely developed as duplications of some common ancestral gene and then diverged in function to confer resistance to different isolates (32). A gene conferring resistance to a related pathogen, *P. polysora*, which causes southern corn rust, mapped near the *Rp1* common rust gene cluster; perhaps, then, it also is a related gene that diverged even further in function to recognize avirulence products of a different species. The clustering of nearly duplicate genes also provides a mechanism (unequal crossing over) for generating the genetic diversity required to create new resistance specificities at a relatively rapid evolutionary rate (32,33).

DNA markers have facilitated the large-scale mapping of many R-genes in many crops, and the results of many studies have revealed that clustering of R-genes is a common phenomenon (although it should be noted that many resistance genes do not exist in clusters). In addition to the *Rp1* cluster in maize, clusters of R-genes affecting a common pathogen (or "complex loci") have been reported in tomato (34), oat (35), barley (36), lettuce (37), and flax (38). Clusters containing genes conferring resistance to different pathogens have also been reported in tomato (39) and *Arabidopsis* (40). A particularly interesting genetic linkage, that of the *Rj2* gene and two fungal disease resistance genes, was reported in soybean (41,42). The *Rj2* gene determines which isolates of *Bradyrhizobia japonica* can nodulate the plant and assist in nitrogen fixation. Although the soybean-*Bradyrhizobium* interaction is a symbiotic one, it nevertheless shares some aspects of plant-pathogen interactions, including specificity of interaction and map position of the specificity gene. Finally, genes that have sequences similar to those of cloned R-genes (see below) also tend to map near known R-genes (43–45).

Taken together, the evidence of genetic clustering supports the hypothesis that genes that confer resistance to different isolates of a common pathogen or even to different pathogens are related and may share some similar functions. Different R-genes seem to be able to trigger a common set of resistance responses; thus, some functional similarities among R-genes are expected. On the other hand, different genes also exhibit variation for specificity; somehow these genes represent different variations on some common functional theme. More direct analysis of R-gene function requires the isolation and sequencing of the genes.

DNA Markers: Map-Based Cloning of Resistance Genes

DNA markers have assisted the analysis of the genetics of disease resistance in one more way, i.e., they have facilitated the isolation of several plant disease R-genes. (Map-based cloning is not the only method to isolate R-genes; transposon tagging has also been used extensively to accomplish this goal.) Map-based cloning of R-genes requires an extremely fine-scale genetic map, and such high resolution can be obtained only by screening very large numbers of progeny (46). For example, in a small sample of segregating progeny, the probability of recovering plants that resulted from a recombination between a particular marker locus and a target gene is very small if the recombination frequency is low. Thus, one may be capable of screening a large number of markers lined to a R-gene, but if the progeny sample size is limited, rare recombinations between the target and marker loci usually will not be observed, and therefore many of the marker loci will be considered to have 0% recombination with the target gene. Unfortunately, the marker loci that cosegregate with the target gene in such a case will cover a rather large physical distance of the chromosome, which may contain hundreds of genes. To identify more recombinations, and therefore to define more precisely the minimum chromosomal area in which the target gene must exist, larger populations must be evaluated both for the resistance phenotype and for DNA marker genotypes.

Once the position of the target gene has been delimited to a reasonably small chromosomal region, libraries of chromosomal fragments from the resistant plant genotype that are propagated clonally in either a yeast or bacterial host can be screened with the markers (46). These libraries represent overlapping fragments of the entire plant genome; thus, the markers defining the target gene area can be used to identify those clones that represent fragments of the target region. The smallest clone or clone sets that span the entire target area can then be sequenced. Obviously, it is critical to delimit as small a chromosomal region as possible to minimize the amount of DNA sequencing that is required to find the gene.

Even when the target region can be precisely defined, it will likely contain several genes. Computational techniques can be used to predict which parts of the sequence represent genes and some possible functions of the gene product (based on comparisons to the growing databases of previously sequenced genes). Nevertheless, it is not trivial to determine which sequence in the target area is the R-gene. In addition to predicted function or similarity to other genes, positive identification of the R-gene may require analysis of gene expression, analysis of gene content of susceptible genotypes, or complementation analyses. Complementation tests are most powerful, but require the ability to insert the candidate sequences into a susceptible host plant using transgenic techniques, and to recover plants that express the resistance transgene adequately. If this is possible, and a candidate gene is identified that confers resistance in an otherwise susceptible genetic background, that candidate gene is almost surely the target R-gene.

DNA Sequences of Resistance Genes: Commonalities and Differences

To date, >20 plant disease resistance genes have been isolated and sequenced (Table 3). The genes cloned confer resistance to fungi, bacteria, viruses, and nematode pathogens, and were isolated from plants in the families Solanaceae, Cruciferae, Compositae, Linaceae, and Gramineae. The publication of these sequences has provided a wonderful opportunity to compare the sequences and predicted functions of genes conferring resistance to different races of a common pathogen and to different pathogens (reviewed in 47–49). Despite their different taxonomic origins and target pathogens, striking similarities exist among these genes, allowing their classification into several categories according to putative function (Table 3).

Recurring motifs of predicted gene products from different R-genes include leucine-rich repeats (LRR), nucleotide binding sites (NBS), protein kinases (PK), leucine zippers (LZ), and regions with similarity to the *Drosophila Toll* and human interleukin-1 genes (TIR), which are involved in signal transduction (Table 3). The most common motif is the LRR, which is predicted to be involved in protein-protein interactions (Table 3) (49). The LRR domain of the *Xa21* protein is predicted to be localized outside the cell membrane, perhaps to interact with avirulence factors before they enter the cell. Other genes have LRR that are predicted to be localized within the cytoplasm of the cell, suggesting that they interact with pathogen products that have already entered the cell (50–52). These localizations and functions are predicted rather than proven, and in some cases, the predictions based on comparison to known sequences have proven to be incorrect (52). For example, the *RPM1* gene product is actually localized to the plasma membrane rather than solubilized in the cytoplasm, as predicted from sequence information (Table 3) (53).

Many other R-genes have NBS (Table 3). NBS protein regions, such as those found in G-proteins, bind ATP or GTP molecules and function in signal transduction pathways (54). Similarly, PK, LZ, or TIR domains (Table 3) may be involved in signal transduction. Most genes have a combination of these motifs. For example, the *Xa21* gene has an extracellular LRR domain, a transmembrane domain, and an intracellular PK domain, a makeup that resembles epidermal growth factor receptors involved in signal transduction in animals (55). The LRR domain of the *Xa21* protein could interact with avirulence products or other elicitors extracellularly, whereupon it could dimerize or undergo conformational changes, activating the PK domain. The PK domain may then phosphorylate some downstream protein, which in turn might initiate a signal cascade, resulting in the resistance response. None of the other sequenced disease R-genes contain all of these functions, however (Table 3). Therefore, the components of disease resistance signal transduction systems must be able to operate in a modular fashion. Perhaps the solely LRR-types of R-genes (e.g., *Cf-9*) interact with avirulence gene products, and the resulting conformational changes could activate separately encoded components of the signal transduction pathway. Progress is beginning to be made in the identification of the genes whose products are also required for the cloned R-genes to function, some of which may represent downstream targets of the signal transduction pathways (52,56,57). Furthermore, avirulence genes from several pathogens have been cloned and are being used to probe the nature of the host-pathogen recognition process (51,58–61). Characterization of these other components of the recognition and signaling systems will likely be required to gain a full understanding of the mechanisms of resistance. Nevertheless, the general outline of disease resistance signaling pathways is fully compatible with Flor's gene-for-gene concept, i.e., host resistance genes are receptors that interact specifically with some product of the pathogen to activate the resistance response. Pathogens can avoid the resistance reaction and cause disease by changing (or losing) their avirulence gene products so that the host does not recognize the pathogen. This also is consistent with the generally observed dominance for both host resistance and pathogen avirulence. Emerging molecular data have also revealed examples of deviations from the strict gene-for-gene hypothesis. For example, the *Arabidopsis* R-gene *Rpm1* confers resistance to pathogens carrying either of two unrelated avirulence genes (62), and the tomato *Mi* gene confers resistance to both nematode and aphid pests (63,64). In addition, the *mlo* gene of barley, which conditions race-nonspecific powdery mildew resistance has been cloned, and it is not related to the gene-for-gene disease resistance genes (65). Studies of this gene may help to understand

TABLE 3
Conserved Sequence Motifs Among Cloned Plant Disease Resistance Genes

Gene(s)	Plant species	Pathogen and disease[b]	Predicted protein motifs[a]					Cellular localization[c]	Reference
			LRR	NBS	LZ	TIR	PK		
Pto	Tomato	*Pseudomonas syringae* (B), bacterial speck					X	C	82
Cf-2, Cf-4, Cf-5, Cf-9	Tomato	*Cladosporum fulvum* (F), leaf mold fungus	X					E, T, C	83–85
PiB	Rice	*Magnaporthe grisea* (F), rice blast	X	X				C	86
Dm3	Lettuce	*Bremia lactucae* (F), downy mildew	X	X				C	87
Rp1	Maize	*Puccinia sorghi* (F), common rust	X	X				C	
Xa1	Rice	*Xanthomonas oryzae* (B), bacterial blight	X	X					89
Xa21	Rice	*Xanthomonas oryzae* (B), bacterial blight	X				X	E, T, C	90

RPP8	Arabidopsis	*Peronospera parasitica* (F), downy mildew	X	X	X		C	91
RPM1, *RPS2*, *RPS5*	Arabidopsis	*Pseudomonas syringae* (B)	X	X	X		C	92–95
I2C	Tomato	*Fusarium oxysporum* (F), vascular wilt	X	X	X		C	96
Mi	Tomato	*Meloidogyne incognita* (N), root knot nematode	X	X	X		C	97
L6, M	Flax	*Melampsorum lini* (F), rust	X	X		X	C	98,99
N	Tobacco	Tobacco mosaic virus (V)	X	X		X	C	100
RPP1, RPP5	*Arabidopsis*	*Peronospera parasitica* (F), downy mildew	X	X		X	C	101,102
RPS4	*Arabidopsis*	*Pseudomonas syringae* (B)	X	X		X	C	103

[a]Abbreviations: LRR, leucine rich-repeat; NBS, nucleotide binding site; LZ, leucine zipper; TIR, region with homology to *Drosophila Toll* gene and human Interleukin-1 gene; PK, protein kinase.
[b]Class of pathogen: B, bacterium; F, fungus; N, nematode; V, virus.
[c]Localization: C, cytoplasmic; E, extracellular; T, transmembrane.

how a single gene can confer broad-spectrum resistance, although it does not appear to be a typical partial resistance gene.

Because LRR are thought to be involved in protein-protein interactions and occur frequently among disease resistance genes, it follows that these are the regions that define specificity of the host-pathogen interaction. If true, we would expect that genes from a common host species that confer resistance to different races of the same pathogen will differ primarily in the LRR regions, and will be structurally similar in other regions. Comparison of sequences of several R-genes that map to the *Dm3* cluster in lettuce and confer resistance to different isolates of the *Bremia lactucae* pathogen bears this out. The different genes are clearly members of a common gene family, but they differ primarily in specific hypervariable LRR regions (48). The only difficulty with the hypothesis that LRR are responsible for direct interaction with pathogen avirulence products is that no such interactions have been observed directly (although they have for the *Pto* protein kinase and its corresponding avirulence gene) (59,60). This lack of evidence does not invalidate the hypothesis, but Martin (52) suggested that LRR genes act downstream of the genes that recognize pathogen avirulence genes or elicitors. More cell biology and biochemistry studies of these systems (rather than exclusively genetic analyses) will be required to identify the physical nature of the host-pathogen interaction that leads to disease resistance.

Once the host R-gene products have interacted with the pathogen and initiated a signal cascade, what are the cellular mechanisms by which resistance is mediated? In gene-for-gene resistances, resistance is largely conferred by hypersensitive responses (47). The hypersensitive response is a rapid cell death that is limited to the infected cell and sometimes a few surrounding cells (47). The hypersensitive response prevents the progression of disease by destroying the tissues upon which the pathogen depends for growth. It is localized cell suicide, wherein the loss of few infected cells contributes to fitness because it maintains the overall health of the whole plant. The cellular processes involved in the hypersensitive response are beginning to be identified, and these include fluxes of Ca^{+2}, nitric oxide, salicylic acid, and superoxide concentrations (47,50,66,67). Rapid progress is expected in unraveling the genes and mechanisms involved in gene-for-gene hypersensitive response resistances. On the other hand, little molecular information exists about genes conferring partial resistance or the mechanisms by which partial resistance is mediated.

Transgenic Strategies to Enhance Disease Resistance

Having isolated genes involved in resistance, identified some of the functional characteristics of these genes, and gained a primitive knowledge of the mechanisms of resistance, how can this knowledge be used to enhance disease resistance in crops? One obvious general strategy is to use genetic transformation technologies to introduce R-genes (modified or otherwise) into susceptible cultivars. The following discussion will be focused on the use of components involved in gene-for-gene resistance mechanisms already described, but it should be noted that other strategies, including pathogen-derived transgenic resistances (used to confer virus resistance) (68,69) and overexpression of antimicrobial compounds (such as chitinase) (70), may prove successful for certain pathogens.

One use of cloned R-genes is to simplify gene pyramiding. Cassettes of multiple R-genes with varying specificities could be introduced into cultivars transgenically to obviate the need for multiple generations of backcrossing and selection for multiple genes (29,71). This also offers the attractive possibility of pyramiding multiple resistance specificities that are allelic in nature. No more than one or two alleles per locus can be maintained in a cultivar of diploid self-pollinating or outcrossing species, respectively. Using artificial gene constructs, however, multiple alleles could be arranged linearly along the same piece of DNA and maintained in a single transgenic pure-line or hybrid cultivar. Furthermore, introduction of different race-specific R-genes into different plants of a common genotype, as suggested by Staskawicz *et al.* (49) could make the development of multiline cultivars more practical (71). Finally, R-genes can be transferred across species barriers, which may be useful if a crop species lacks resistance to a disease for which R-genes have been isolated in other species (29,71). The implementation of these strategies will require the isolation of multiple members of disease resistance families. The technical difficulties of avoiding gene silencing (particularly with multiple gene constructs) and engineering the proper timing and quantity of gene expression that face all transformation projects will also have to be overcome.

Knowledge of the critical sites within hypervariable regions of receptors that define the specificity of resistance could be exploited by developing R-genes with novel specificities or by designing genes that will recognize characteristics of the pathogen that are unique, but critical for pathogen survival (29,71). The latter approach might reduce the possibility that the pathogen can easily overcome the resistance by mutation or loss of the detected characteristic.

Introduction of transgenes that cause resistance responses to be activated constitutively has been proposed (49), but it is not clear how the generally detrimental effects of the resulting constitutive defense responses could be avoided. The tight regulation of the hypersensitive response and other disease responses has evolved surely because it confers greater fitness than constitutive expression.

Directed Gene Modification

There is a major disadvantage to transgenic approaches to enhancing disease resistance, i.e., the opposition to genetically modified organisms in many countries and by many consumers. Currently, there is no other practical method to introduce novel genes into crops, but some newly emerging technologies may be useful someday for enhancing disease resistance or any other crop attribute by allowing directed gene sequence modification without introduction of alien DNA into the modified genotype. Chimeraplasty, or chimeric oligonucleotide–dependent mismatch repair, is one such technology that has been adapted recently from mammalian systems for use in plants (72). Briefly, the method is able to introduce 1- to 3-bp alterations into a target gene sequence by the introduction of oligonucleotides containing the desired altered sequence and ~10 bp flanking either side of this sequence that are homologous to the target gene region. The oligonucleotide is composed of a single strand of nucleotides, including complementary stretches of the target gene. These complementary bases cause the molecule to anneal with itself; one complementary set is composed of DNA bases, whereas the other contains DNA and methylated RNA bases. When this chimeric oligonucleotide (CO) is introduced into the plant nucleus (using the same methods developed for introducing transgenes *via* particle bombardment), it associates specifically with the homologous target region. The gap between the first and last bases of the CO molecule may be sensed as a break in the chromosome by the cell's DNA repair mechanisms. Presumably, the endogenous DNA mismatch repair mechanisms "correct" the difference between the CO and plant DNA sequences by modifying the plant sequence to match the target sequence of the CO, including its one or few modified bases (72). The result is a change in the plant's target gene that matches the introduced oligonucleotide, which is heritable and involves no introduction of foreign DNA (73). Unlike transgenic approaches, the oligonucleotide is not incorporated into the chromosome. This method has been used successfully to modify genes in maize and tobacco (73,74). Unfortunately, the frequency of desired mutations that are recovered is on the order of 10^{-4} at best. This frequency is too low to permit practical application of this method for general use (72). Furthermore, unexpected changes are sometimes observed at the target site (73,74). But if the method does eventually become practicable, it could be used to introduce specific short modifications into the hypervariable regions of disease resistance gene LRR, thereby modifying the specificity of the interaction in one of the ways already outlined.

Durability of Transgenic or Chimeraplasty-Enhanced Resistances

The question remains: Is there any reason to be optimistic that transgenically derived sources of disease resistance will be more durable than natural gene-for-gene resistances? Insofar as transgenes mimic the mode of action of, and are deployed in the same manner as natural hypersensitive resistance genes, the answer is no. The durability of even multiple R-gene cassettes must remain suspect for the same reasons that natural gene pyramids are not necessarily expected to be durable. Attempts to develop new race specificities *via* directed gene modifications of LRR regions of R-genes may turn out to be an extremely expensive method of creating nondurable resistances. Finally, the widespread deployment of common resistance transgenes in multiple crops may hasten the breakdown of resistance (29).

Perhaps R-genes can be developed by rational design to recognize components that are vital to the pathogen (29). In this case, the fitness cost to the pathogen of mutations required to avoid recognition may be so high as to prevent the development of virulence against plants carrying such a gene. It is not obvious, however, which components are necessary for pathogen survival, nor is it obvious how such components could be detected consistently and specifically upon infection. Why have most putative receptor-type R-genes evolved to have cytoplasmic rather than extracellular localizations (Table 3)? Is it because specific recognition of pathogen components is more difficult outside than inside the cell membrane? If this is true, it makes the recognition of vital pathogen attributes more difficult because most if not all vital pathogen components will never enter the plant cell membrane. Furthermore, if the pathogen is not detected with very high specificity, "false positive" recognitions would result in hypersensitive responses that will damage the host. More fundamentally, the capacity for many pathogen populations to rapidly evolve mechanisms that will permit their survival when agricultural systems subject them to strong selection pressures should not be underestimated. More potential evolutionary plasticity may exist in even seemingly vital pathogen components than might be suspected. Because evolution relies in part on chance mutations, it is a numbers game. The grand evolutionary experiment undertaken by plant breeders in the 20th century demonstrated that the numbers favor microbial pathogens, with their astronomical reproductive capacity, over plant breeders, whose ability to generate and manage breeding lines is much more limited (5). There is little reason to believe that even the most elegantly conceived resistance mechanisms will remain invincible when deployed on a large scale.

General Strategies for Disease Management and Prevention

A total systems approach is the best hope for sustainable management of pests and diseases of agricultural crops. The current disease management paradigm seeks "therapeutic" or "silver bullet" solutions that will eliminate a pathogen. In contrast, a total system approach emphasizes the promotion of natural checks against disease that are components (or perhaps emergent properties) of agricultural ecosystems and that help to bring the pathogens into balance within the system, rather than eliminate them (75). Disease is not simply caused by a pathogen; rather, it is the result of an interaction among host, pathogen, and the environment. The environment includes the agroecosystem, and that part of the environment, at least, is liable to human modification.

Diversity of crop genotypes can help to stabilize pathogen populations. For example, wild oat (*A. sterilis*) populations have existed in balance with crown rust fungus (*Puccinia coronata*) populations since before the development of agriculture in the Middle East. Many resistance genes or alleles exist in the host populations, but none at high frequency (76). In contrast, as soon as single R-genes began to be deployed on a large scale by humans in North America, an unstable boom-and-bust cycle of resistance and susceptibility immediately resulted (77). As discussed previously, multiline cultivars and cultivar mixtures offer practical alternatives to monogenotypic monocultures, and the benefits of these alternatives for disease control have been documented (77,78).

Species diversity in agroecosystems (introduced *via* crop rotations and intercropping) can also help to minimize disease problems. For example, head scab of barley and wheat (*Fusarium graminearum*) recently emerged as a major problem in the cereal growing regions of the North Central United States. The response to this outbreak has been a scramble to identify resistance genes and to incorporate them into commercially acceptable cultivars. Changes in crop rotations in these regions could easily minimize the problem. For example, oat could be grown as an alternative cereal crop because it never develops the disease. Research priorities are generally not given to development of appropriate agroecosystems and crop rotations or to improvement of alternative crops, however, because in the current paradigm of specialized cropping systems, the solution must be found within the wheat or barley itself. Similarly, millions of research dollars are spent in attempts to map and clone genes for soybean cyst nematode (*Heterodera glycines*) resistance, despite the fact that the disease can be managed easily by growing soybeans in rotations that include 2 y of non-host crops (79). From the total system perspective, the disease is not the primary problem; rather it is the result of the loss of diversity in the agroecosystem. Many solutions to agricultural problems have been developed in disregard of their effects on the whole agricultural system, and, when adapted, have become the source of new problems (80).

Within a total system approach to disease management, the tools of biotechnology may seem less urgent:

> "Transgenic plants may provide some worthwhile advance on the already huge improvements in resistance achieved by conventional plant breeding, but a much greater contribution to food security would be made by more sustainable management of the soil. When we ask what biotechnology can contribute to disease control in farming systems which rely on genetic and species diversity, allied to careful management of the soil and other resources, the answer may be 'nothing at all.' In these situations, no one gene (or even any one crop) offers a substantial benefit by itself" (81).

At the same time, deployment of biotechnologically or naturally derived resistances within a total system approach will improve their durability. The total system approach will incorporate other checks against disease and also permit survival of the pathogen in the system, thus reducing selection pressure for new virulence against these expensively developed forms of resistance. "Crops engineered to express toxins of pathogens are simply targeted as replacements for synthetic pesticides and will become ineffective in the same way pesticides have. It will be unfortunate if these powerful agents are wasted rather than integrated as key parts of sustainable pest management systems" (75). Thus, rather than choosing between ecological management approaches or plant breeding and biotechnology, we can and should use both.

Acknowledgment

I thank Dr. Jim Kolmer for helpful comments on a previous draft of this manuscript.

References

1. Flor, H.H., Host-Parasite Interaction in Flax-Rust—Its Genetics and Other Implications, *Phytopathology 43:* 624–628 (1955).
2. Flor, H.H., The Complementary Genic Systems in Flax and Flax Rust, *Adv. Genet. 8:* 29–54 (1956).
3. Flor, H.H., Current Status of the Gene-for-Gene Concept, *Annu. Rev. Phytopathol. 9:* 275–296 (1971).
4. Vanderplank, J.E., *Disease Resistance in Plants,* 2nd ed., Academic Press, Orlando, 1984.
5. Johnson, T., Man-Guided Evolution in Plant Rusts, *Science 133:* 357–362 (1961).
6. Frey, K.J., Oat Improvement with Genes from Avena species, in *Proceedings of the 4th International Oat Conference, Vol. 2,* edited by A.R. Barr, and R.W. Medd, Adelaide, Australia, 1992, pp. 61–64.
7. Holland, J.B., Oat Improvement, in *Crop Improvement for the 21st Century,* edited by M.S. Kang, Research Signpost, Trivandrum, India, 1997, pp. 57–98.
8. Kolmer, J.A., Virulence Dynamics and Genetics of Cereal Rust Populations in North America, in *The Gene-for-Gene Relationship in Plant-Parasite Interactions,* edited by I.R. Crute, E.B. Holub, and J.J. Burdon, CAB International, New York, 1997, pp. 139–156.
9. Brown, J.K.M., E.M. Foster, and R.B. O'Hara, Adaptation of Powdery Mildew Populations to Cereal Varieties in Relation to Durable and Non-Durable Resistance, in *The Gene-for-Gene Relationship in Plant Parasite Interactions*, edited by I.R. Crute, E.B. Holub, and J.J. Burdon, CAB International, New York, 1997, pp. 119–138.
10. Parlevliet, J.E., What Is Durable Resistance, a General Outline, in *Durability of Disease Resistance,* edited by T. Jacobs, and J.A. Parlevliet, Kluwer, Dordrecht, 1992, pp. 23–39.
11. Burdon, J.J., Genetic Variation in Pathogen Populations and Its Implications for Adaptation to Host Resistance, in *Durability of Disease Resistance,* edited by T. Jacobs, and J.A. Parlevliet, Kluwer, Dordrecht, 1992, pp. 41–56.
12. Chong, J., and J.A. Kolmer, Virulence Dynamics and Phenotypic Diversity of *Puccinia coronata* f. sp. *avenae* in Canada from 1974 to 1990, *Can. J. Bot. 71:* 248–255 (1993).
13. Johnson, R., Durable Resistance: Definition of, Genetic Control, and Attainment in Plant Breeding, *Phytopathology 71:* 567–568 (1981).
14. Finkh, M.R., and C.C. Mundt, Stripe Rust, Yield, and Plant Competition in Wheat Cultivar Mixtures, *Phytopathology 82:* 905–913 (1992).
15. Newton, A.C., Cultivar Mixtures in Intensive Agriculture, in *The Gene-for-Gene Relationship in Plant-Parasite Interactions,* edited by I.R. Crute, E.B. Holub, and J.J. Burdon, CAB International, New York, 1997, pp. 65–80.
16. Wolfe, M.S., J.A. Barret, and J.E.E. Jenkins, The Use of Cultivar Mixtures for Disease Control, in *Strategies for the Control of Cereal Disease,* edited by J.F. Jenkyn, and R.T. Plumb, Blackwell, Oxford, 1981, pp. 73–80.

17. Browning, J.A., M.D. Simons, and E. Torres, Managing Host Genes: Epidemiological and Genetic Concepts, in *Plant Disease: An Advanced Treatise,* edited by J.G. Horsfall, and E.B. Cowling, Academic Press, New York, 1977, vol. I, pp. 191–212.
18. Fehr, W.R., *Principles of Cultivar Development. Volume 1: Theory and Technique*, Macmillan Publishing, New York, 1987.
19. Simons, M.D., Polygenic Resistance to Plant Disease and Its Use in Breeding Resistant Cultivars, *J. Environ. Qual. 1:* 232–240 (1972).
20. Johnson, R., Durable Disease Resistance, in *Strategies for the Control of Cereal Disease,* edited by J.F. Jenkyn, and R.T. Plumb, Blackwell, Oxford, 1981, pp. 55–63.
21. Shaner, G., G. Buechley, and W.E. Nyquist, Inheritance of Latent-Period of *Puccinia recondita* in Wheat, Crop Sci. 37: 748–756 (1997).
22. Stuthman, D.D., Oat Breeding and Genetics, in *The Oat Crop: Production and Utilization,* edited by R.W. Welch, Chapman and Hall, London, 1995, pp. 150–176.
23. Brake, V.M., and J.A.G. Irwin, Partial Resistance of Oats to *P. coronata* f. sp. *avenae., Aust. J. Agric. Res. 43:* 1217–1227 (1992).
24. Parlevliet, J.E., Components of Resistance That Reduce the Rate of Epidemic Development, *Euphytica 17:* 203–222 (1979).
25. Kolmer, J.A., Genetics of Resistance to Leaf Rust, *Annu. Rev. Phytopathol. 34:* 435–455 (1996).
26. Holland, J.B., D.V. Uhr, D. Jeffers, and M.M. Goodman, Inheritance of Resistance to Southern Corn Rust in Tropical-by-Corn Belt Maize Populations, *Theor. Appl. Genet. 96:* 232–241 (1998).
27. McMullen, M.D., and K.D. Simcox, Genomic Organization of Disease and Insect Resistance Genes in Maize., *Mol. Plant-Microbe Interact. 8:* 811–815 (1995).
28. Lee, M., DNA Markers and Plant Breeding Programs, *Adv. Agron. 55:* 265–344 (1995).
29. Michelmore, R., Molecular Approaches to Manipulation of Disease Resistance Genes, *Annu. Rev. Phytopathol. 15:* 393–427 (1995).
30. Young, N.D., QTL Mapping and Quantitative Disease Resistance in Plants, *Annu. Rev. Phytopathol. 34:* 479–501 (1996).
31. Cox, T.S., Simultaneous Selection for Major and Minor Resistance Genes, *Crop Sci. 35:* 1337–1346 (1995).
32. Saxena, K.M.S., and A.L. Hooker, On the Structure of a Gene for Disease Resistance in Maize, *Genetics 61:* 1300–1305 (1968).
33. Richter, T.E., T.J. Pryor, J.L. Bennetzen, and S.H. Hulbert, New Rust Resistance Specificities Associated with Recombination in the *Rp1 Complex* in Maize, *Genetics 141:* 373–381 (1995).
34. Jones, D.A., M.J. Dickinson, P.J. Balint-Kurti, M.S. Dixon, and J.D.G. Jones, Two Complex Resistance Loci Revealed in Tomato by Classical and RFLP Mapping of the *Cf-2, Cf-4, Cf-5,* and *Cf-9* Genes for Resistance to *Cladosporum fulvum, Mol. Plant-Microbe Interact. 6:* 348–357 (1993).
35. Wise, R.P., M. Lee, and P.J. Rayapati, Recombination Within a 5-Centimorgan Region in Diploid *Avena* Reveals Multiple Specificities Conferring Resistance to *Puccinia coronata, Phytopathology 86:* 340–346 (1996).
36. Mahadevappa, M., R.A. DeScenzo, and R.P. Wise, Recombination of Alleles Conferring Specific Resistance to Powdery Mildew at the *Mla* Locus in Barley, *Genome 37:* 460–468 (1994).

37. Witsenboer, H., R.V. Kesseli, M.G. Fortin, M. Stanghellini, and R.W. Michelmore, Sources and Genetic Structure of a Cluster of Genes for Resistance to Three Pathogens in Lettuce, *Theor. Appl. Genet. 91:* 178–188 (1995).
38. Ellis, J.G., G.J. Lawrence, E.J. Finnegan, and P.A. Anderson, Contrasting Complexity of Two Rust Resistance Loci in Flax, *Proc. Natl. Acad. Sci. USA 92:* 4185–4188 (1995).
39. Dickinson, M.J., D.A. Jones, and J.D.G. Jones, Close Linkage Between *Cf-2/Cf-5* and *Mi* Resistance Loci in Tomato, *Mol. Plant-Microbe Interact. 6:* 341–347 (1993).
40. Holub, E.B., Organization of Resistance Genes in *Arabidopsis, in The Gene-for-Gene Relationship in Plant-Parasite Interactions,* edited by I.R. Crute, E.B. Holub, and J.J. Burdon, CAB International, New York, 1997, pp. 5–26.
41. Devine, T.E., T.C. Kilen, and J.J. O'Neill, Genetic Linkage of the Phytopthora Resistance Gene *Rps2* and the Nodulation Response Gene *Rj2* in Soybean, *Crop Sci. 31:* 713–715 (1991).
42. Lohnes, D.G., R.E. Wagner, and R.L. Bernard, Soybean Genes *Rj2, Rmd,* and *Rps2* in Linkage Group 19, *J. Hered. 84:* 109–111 (1993).
43. Kanazin, V., L.F. Marek, and R.C. Shoemaker, Resistance Gene Analogs Are Conserved and Clustered in Soybean, *Proc. Natl. Acad. Sci. USA 93:* 11746–11750 (1996).
44. Leister, D., J. Kurth, D.A. Laurie, M. Yano, T. Sasaki, K. Devos, A. Graner, and P. Schulze-Lefert, Rapid Reorganization of Resistance Gene Homologues in Cereal Genomes, *Proc. Natl. Acad. Sci. USA 95:* 370–375 (1998).
45. Shen, K.A., B.C. Meyers, M.N. Islam-Faridi, D.B. Chin, D.M. Stelly, and R.W. Michelmore, Resistance Gene Candidates Identified by PCR with Degenerate Oligonucleotide Primers Map to Clusters of Resistance Genes in Lettuce, *Mol. Plant-Microbe Interact. 11:* 815–823 (1998).
46. Tanksley, S.D., M.W. Ganal, and G.B. Martin, Chromosome Landing: A Paradigm for Map-Based Gene Cloning in Plants with Large Genomes, *Trends Genet. 11:* 63–68 (1995).
47. Hammond-Kosack, K.E., and J.D.G. Jones, Resistance Gene-Dependent Plant Defense Responses, *Plant Cell 8:* 1773–1791 (1996).
48. Michelmore, R.W., and B.C. Meyers, Clusters of Resistance Genes in Plants Evolve by Divergent Selection and Birth-and-Death Process, *Genome Research 8:* 1113–1130 (1998).
49. Staskawicz, B.J., F.M. Ausubel, B.J. Baker, J.G. Ellis, and J.D.G. Jones, Molecular Genetics of Plant Disease Resistance, *Science 268:* 661–667 (1995).
50. Grant, M., and J. Mansfield, Early events in host-pathogen interactions, *Curr. Opin. Plant Biol. 2:* 312–319 (1999).
51. Holt, B.F., D. Mackey, and J.L. Dangl, Primer—Recognition of Pathogens by Plants, *Curr. Biol. 10:* R5–R7 (2000).
52. Martin, G.B., Functional Analysis of Plant Disease Resistance Genes and Their Downstream Effectors, *Curr. Opin. Plant Biol. 2:* 273–279 (1999).
53. Boyes, D.C., J. Nam, and J. Dangl, The *Arabidopsis thaliana* RPM1 Disease Resistance Gene Product Is a Peripheral Plasma Membrane Protein That Is Degraded Coincident with the Hypersensitive Response, *Proc. Natl. Acad. Sci. USA 95:* 15849–15854 (1998).
54. Rensdomiano, S., and H.E. Hamm, Structural and Functional-Relationships of Heterotrimeric G-Proteins, *FASEB J. 9:* 1059–1066 (1995).

55. Lewin, B., *Genes VI*, Oxford University Press, Oxford, 1997.
56. Glazebrook, J., Genes Controlling Expression of Defense Responses in Arabidopsis, *Curr. Opin. Plant Biol. 2:* 280–286 (1999).
57. Piffanelli, P., A. Devoto, and P. Schulze-Lefert, Defence Signalling Pathways in Cereals, *Curr. Opin. Plant Biol. 2:* 295–300 (1999).
58. Knogge, W., and C. Marie, Molecular Characterization of Fungal Avirulence, in *The Gene-for-Gene Relationship in Plant-Parasite Interactions,* edited by I.R. Crute, E.B. Holub, and J.J. Burdon, CAB International, New York, 1997, pp. 329–346.
59. Schofield, S.R., C.M. Tobias, J.P. Rathjen, J.H. Chang, D.T. Lavelle, R.W. Michelmore, and B.J. Staskawicz, Molecular Basis of Gene-for-Gene Specificity in Bacterial Speck Disease of Tomato, *Science 274:* 2063–2065 (1996).
60. Tang, X., R.D. Frederick, J. Zhou, D.A. Halterman, Y. JIa, and G.B. Martin, Initiation of Plant Disease Resistance by Physical Interaction of AvrPto and Pto Kinase, *Science 274:* 2060–2063 (1996).
61. Vivian, A., M.J. Gibbon, and J. Murillo, The Molecular Genetics of Specificity Determinants in Plant Pathogenic Bacteria, in *The Gene-for-Gene Relationship in Plant-Parasite Interactions,* edited by I.R. Crute, E.B. Holub, and J.J. Burdon, CAB International, New York, 1997, pp. 293–328.
62. Bisgrove, S.R., M.T. Simonich, N.M. Smith, R.W. Sattler, and R.W. Innes, A Disease Resistance Gene in *Arabidopsis* with Specificity for Two Different Pathogen Avirulence Genes, *Plant Cell 6:* 927–933 (1994).
63. Rossi, M., F.L. Goggin, S.B. Milligan, I. Kaloshian, D.E. Ullman, and V.M. Williamson, the Nematode Resistance Gene Mi of Tomato Confers Resistance Against the Potato Aphid, *Proc. Natl. Acad. Sci. USA 95:* 9750–9754 (1998).
64. Vos, P., G. Simons, T. Jesse, J. Wijbrandi, L. Heinen, R. Hogers, A. Frijters, J. Groenendijk, P. Diergaarde, M. Reijans, J. Fierens-Onstenk, M. de Both, J. Peleman, T. Liharska, J. Hontelez, and M. Zabeau, The Tomato Mi-1 Gene Confers Resistance to Both Root-Knot Nematodes and Potato Aphids, *Nat. Biotechnol. 16:* 1365–1369 (1998).
65. Büschges, R., K. Hollricher, R. Panstruga, G. Simons, M. Wolter, A. Frijters, R. van Daelen, T. van der Lee, P. Diergaarde, J. Groenendijk, S. Töpsch, P. Vos, F. Salamini, and P. Schulze-Lefert, The Barley *Mlo* Gene: A Novel Control Element of Plant Pathogenic Resistance, *Cell 88:* 695–705 (1997).
66. Delledonne, M., Y. Xia, R.A. Dixon, and C. Lamb, Nitric Oxide Functions as a Signal in Plant Disease Resistance, *Nature 394:* 585–527 (1998).
67. McDowell, J.M., and J.L. Dangl, Signal Transduction in the Plant Immune Response, *Trends Biochem. Sci. 25:* 79–82 (2000).
68. Fitchen, J.H., and R.N. Beechy, Genetically Engineered Protection Against Viruses in Transgenic Plants, *Annu. Rev. Microbiol. 47:* 739–763 (1993).
69. Lomonossoff, G.P., Pathogen-Derived Resistance to Plant Viruses, *Annu. Rev. Phytopathol. 33:* 323–343 (1995).
70. Zhu, Q., E.A. Maher, S. Masoud, R.A. Dixon, and C.J. Lamb, Enhanced Protection Against Fungal Attack by Constitutive Co-Expression of Chitinase and Glucanase Genes in Transgenic Tobacco, *Bio/Techniques 12:* 807–812 (1994).
71. Crute, I.R., The Elucidation and Exploitation of Gene-for-Gene Recognition, *Plant Pathol. 47:* 107–113 (1998).
72. Hohn, B., and H. Puchta, Gene Therapy in Plants, *Proc. Natl. Acad. Sci. USA 96:* 8321–8323 (1999).

73. Zhu, T., D.J. Peterson, L. Tagliani, G. St Clair, C.L. Baszczynski, and B. Bowen, Targeted Manipulation of Maize Genes *In Vivo* Using Chimeric RNA/DNA Oligonucleotides, *Proc. Natl. Acad. Sci. USA 96:* 8768–8773 (1999).
74. Beetham, P.R., P.B. Kipp, X.L. Sawycky, C.J. Arntzen, and G.D. May, A Tool for Functional Plant Genomics: Chimeric RNA/DNA Oligonucleotides Cause *In Vivo* Gene-Specific Mutations, *Proc. Natl. Acad. Sci. USA 96:* 8774–8778 (1999).
75. Lewis, W.J., J.C. van Lenteren, S.C. Phatak, and J.H. Tumlinson, III, A Total System Approach to Sustainable Pest Management, *Proc. Natl. Acad. Sci. USA 94:* 12243–12248 (1997).
76. Segal, A., J. Manisterski, G. Fischbeck, and I. Wahl, How Plant Populations Defend Themselves in Natural Ecosystems, in *Plant Disease: An Advanced Treatise. Vol. V: How Plants Defend Themselves,* edited by J.G. Horsfall, and E.B. Cowling, Academic Press, New York, 1980, pp. 75–102.
77. Browning, J.A., One Phytopathologist's Growth Through IPM to Holistic Plant Health: The Key to Approaching Genetic Yield Potential, *Annu. Rev. Phytopathol. 36:* 1–24 (1998).
78. Smithson, J.B., and J.M. Lenné, Varietal Mixtures: A Viable Strategy for Sustainable Productivity in Subsistence Agriculture, *Ann. Appl. Biol. 128:* 127–158 (1996).
79. Wrather, M.P., S.C. Anand, and S.R. Koenning, Management by Cultural Practices, in *Biology and Management of the Soybean Cyst Nematode,* edited by R.D. Riggs, and J.A. Wrather, ASA Press, St. Paul, 1992, pp. 125–133.
80. Berry, W., *The Gift of Good Land,* North Point Press, San Francisco, 1980.
81. Brown, J.K.M., How to Feed the World, in Two Contradictory Lessons, *Trends Plant Sci. 3:* 409–410 (1998).
82. Martin, G.B., S.H. Brommonschenkel, J. Chunwongse, A. Frary, M.W. Ganal, R. Spivey, T. Wu, E.D. Earle, and S.D. Tanksley, Map-Based Cloning of a Protein Kinase Gene Conferring Disease Resistance in Tomato, *Science 262:* 1432–1436 (1993).
83. Dixon, M.S., D.A. Jones, J.S. Keddie, C.M. Thomas, K. Harrison, and J.D.G. Jones, The Tomato *Cf-2* Disease Resistance Locus Comprises Two Functional Genes Encoding Leucine- Rich Repeat Proteins, *Cell 84:* 451–459 (1996).
84. Hammond-Kosack, K.E., and J.D.G. Jones, Plant Disease Resistance Genes, *Annu. Rev. Plant Physiol. Plant Mol. Biol. 48:* 575–607 (1997).
85. Jones, D.A., C.M. Thomas, K.E. Hammond-Kosack, P.J. Balint-Kurti, and J.D.G. Jones, Isolation of the Tomato *Cf-9* Gene for Resistance to *Cladosporum fulvum* by Transposition Tagging, *Science 266:* 789–793 (1994).
86. Wang, Z.X., M. Yano, U. Yamanouchi, M. Iwamoto, L. Monna, H. Hayasaka, Y. Katayose, and T. Sasaki, The *Pib* Gene for Rice Blast Resistance Belongs to the Nucleotide Binding and Leucine-Rich Repeat Class of Plant Disease Resistance Genes, *Plant J. 19:* 55–64 (1999).
87. Meyers, B.C., K.A. Shen, P. Rohani, B.S. Gaut, and R.W. Michelmore, Receptor-Like Genes in the Major Resistance Locus of Lettuce Are Subject to Divergent Selection, *Plant Cell 10:* 1833–1846 (1998).
88. Collins, N.C., J. Drake, M. Aycliffe, Q. Sun, J. Ellis, S. Hulbert, and A. Pryor, Molecular Characterization of the Maize *Rp1-D* Rust Resistance Haplotype and Its Mutants, *Plant Cell 11:* 1365–1376 (1999).
89. Yoshimura, S., U. Yamanouchi, Y. Katayose, S. Toki, Z.-X. Wang, I. Kono, N. Kurata, M. Yano, N. Iwata, and T. Sasaki, Expression of *Xa1*, a Bacterial Blight-

Resistance Gene in Rice, Is Induced by Bacterial Inoculation, *Proc. Natl. Acad. Sci. USA 95:* 1663–1668 (1998).

90. Song, W.-Y., G.-L. Wang, L.-L. Chen, H.-S. Kim, L.-Y. Pi, T. Holsten, J. Gardner, B. Wang, W.X. Zhai, L.-H. Zhu, C. Fauquet, and P. Ronald, A Receptor Kinase-Like Protein Encoded by the Rice Disease Resistance Gene, *Xa21, Science 270:* 1804–1806 (1995).
91. McDowell, J.M., M. Dhandaydham, T.A. Long, M.G.M. Aarts, S. Goff, E.B. Holub, and J.L. Dangl, Intragenic Recombination and Diversifying Selection Contribute to the Evolution of Downy Mildew Resistance at the RPP8 Locus of Arabidopsis, *Plant Cell 10:* 1861–1874 (1998).
92. Bent, A.F., B.N. Kunkel, D. Dahlbeck, K.L. Brown, R. Schmidt, J. Giraudat, J. Leung, and B.J. Staskawicz, *RPS2* of *Arabidopsis thaliana:* A Leucine-Rich Repeat Class of Plant Disease Resistance Genes, *Science 265:* 1856–1860 (1994).
93. Grant, M.R., L. Godiard, E. Straube, T. Ashfield, J. Lewald, A. Sattler, R.W. Innes, and J.L. Dangl, Structure of the *Arabidopsis Rpm1* Gene Enabling Dual-Specificity Disease Resistance, *Science 269:* 843–846 (1995).
94. Mindrinos, M., F. Katagiri, G.-L. Yu, and F.M. Susubel, The *A. thaliana* Disease Resistance Gene *RPS2* Encodes a Protein Containing a Nucleotide-Binding Site and Leucine- Rich Repeats, *Cell 78:* 1089–1099 (1994).
95. Warren, R.F., A. Henk, P. Mowery, E. Holub, and R.W. Innes, A Mutation Within the Leucine-Rich Repeat Domain of the *Arabidopsis* Disease Resistance Gene *RPS5* Partially Suppresses Multiple Bacterial and Downy Mildew Resistance Genes, *Plant Cell 10:* 1439–1452 (1998).
96. Ori, N., Y. Eshed, I. Paran, G. Presting, D. Aviv, S. Tanksley, D. Zamir, and R. Fluhr, The *I2C* Family from the Wilt Disease Resistance Locus I2 Belongs to the Nucleotide Binding, Leucine-Rich Repeat Superfamily of Plant Resistance Genes, *Plant Cell 9:* 521–532 (1997).
97. Milligan, S.B., J. Bodeau, J. Yaghoobi, I. Kaloshian, P. Zabel, and V.M. Williamson, The Root Knot Nematode Resistance Gene *Mi* from Tomato Is a Member of the Leucine Zipper, Nucleotide Binding, Leucine-Rich Repeat Family of Plant Genes, *Plant Cell 10:* 1307–1319 (1998).
98. Anderson, P.A., G.J. Lawrence, B.C. Morrish, M.A. Ayliffe, E.J. Finnegan, and J.G. Ellis, Inactivation of the Flax Rust Resistance Gene *M* Associated with Loss of a Repeated Unit Within the Leucine-Rich Repeat Coding Region, *Plant Cell 9:* 641–651 (1997).
99. Lawrence, G.J., E.J. Finnegan, M.A. Ayliffe, and J.G. Ellis, The *L6* Gene for Flax Rust Resistance Is Related to the *Arabidopsis* Bacterial Resistance Gene *RPS2* and the Tobacco Viral Resistance Gene *N, Plant Cell 7:* 1195–1206 (1995).
100. Whitham, S., S.P. Dinesh-Kumar, D. Choi, R. Hehl, C. Corr, and B. Baker, The Product of the Tobacco Mosaic Virus Resistance Gene *N*: Similarity to Toll and the Interleukin-1 Receptor, *Cell 78:* 1101–1115 (1994).
101. Botella, M.A., J.E. Parker, L.N. Frost, P.D. Bittner-Eddy, J.L. Beynon, M.J. Daniels, E.B. Holub, and J.D.G. Jones, Three Genes of the *Arabidopsis RPP1* Complex Resistance Locus Recognize Distinct *Peronospora parasitica* Avirulence Determinants, *Plant Cell 10:* 1847–1860 (1998).
102. Parker, J.E., M.J. Coleman, V. Szabo, L.N. Frost, R. Schmidt, E.A. vanderBiezen, T. Moores, C. Dean, M.J. Daniels, and J.D.G. Jones, The *Arabidopsis* Downy Mildew

Resistance Gene *RPP5* Shares Similarity to the Toll and Interleukin-1 Receptors with *N* and *L6, Plant Cell 9:* 879–894 (1997).

103. Gassmann, W., M.E. Hinsch, and B.J. Staskawicz, The *Arabidopsis RPS4* Bacterial-Resistance Gene Is a Member of the TIR-NBS-LRR Family of Disease-Resistance Genes, *Plant J. 20:* 265–277 (1999).

Chapter 10

Structure, Function, and Engineering of Microbial and Plant β-Ketoacyl-Acyl Carrier Protein Synthase III

Amine Abbadi and Friedrich Spener

Department of Biochemistry, University of Münster, 48149 Münster, Germany

Introduction

Fatty acid biosynthesis occurs through sequential additions of malonyl-derived two-carbon units to acyl acceptors. The addition requires condensation, a first reduction, dehydration, and a second reduction in each cycle. In some bacteria and in plants, each step of this biosynthesis is catalyzed by type II fatty acid synthase (FAS type II), which is composed of individual polypeptides. The β-ketoacyl-acyl carrier protein synthase (KAS), commonly referred to as a condensing enzyme, catalyzes the crucial chain-lengthening C-C bond formation (1,2). In contrast to multifunctional FAS type I of animals and yeast, there are to date four different known FAS type II condensing enzymes (3–8). The first discovery of KAS III in bacteria (5) and in plants (6) gave new insights into the starting reaction of fatty acid biosynthesis in these organisms. This enzyme initiates the chain of reactions by catalyzing the condensation of acetyl-CoA with malonyl-acyl carrier protein (ACP) to produce β-ketobutyryl-ACP, which, on completion of the first cycle of fatty acid synthase reaction, is reduced to butyryl-ACP, using pyrimidine nucleotides as reducing equivalents. The use of a CoA-derivative instead of an ACP-derivative as the primer makes KAS III unique in view of other condensing enzymes, namely, KAS I, KAS II, and KAS IV. On the basis of the pioneering work of Shimakata and Stumpf (3,4,9) KAS I is considered to be responsible for chain elongations up to myristinoyl-ACP, and KAS II for further elongation to palmitoyl- and stearoyl-ACP. The newly discovered KAS IV seems to have a strict specificity for short- and medium-chain acyl-ACP (7,8,10,11).

In this chapter, we review the unique properties of KAS III because recent progress in understanding their structure, mechanisms, and regulation may furnish fundamental knowledge concerning condensing enzymes in bacteria and plants in general, and may lead to potential biotechnological applications of KAS III in particular.

Characterization

Genes and cDNAs

During the last decade, great progress has been made in cloning KAS III cDNAs. But enzymes for which cDNA has become available remain limited to a few organisms.

Initially, a KAS III gene of 951 bp was obtained from *Escherichia coli* and named *fabH* (12). Using this information, two open reading frames (ORF), *yjaX* with 971 bp and *yhfb* with 973 bp, were recently identified in gram-positive *Bacillus subtilis*, which encode FAbH1 and FabH2, isoforms of *B. subtilis* KAS III, respectively (13).

The first report of a KAS III gene from any plant species, as well as the first report of an enzyme involved in fatty acid biosynthesis and encoded by the chloroplast genome, was provided by Reith (14). In the process of sequencing the chloroplast genome of the red alga *Porphyra umbilicalis* this author detected an ORF showing significant homology not only to *fabH* from *E. coli*, but also to an ORF located upstream of *Rhodobacter capsulatus himA* and to a partial ORF located upstream of *Pseudomonas aeruginosa phnA*. On the basis of these homologies, the *P. umbilicalis* KAS III gene also has been named *fabH*.

With the aim of changing KAS III levels in plants, efforts in cloning KAS III genes were concentrated largely on dicotyledonous plants. On the basis of sequence information derived from KAS III purified from spinach leaves (*Spinacea oleracea*), the screening of spinach leaf and root cDNA libraries resulted in the identification of an ORF of 1218 bp (15). In addition, a cDNA library from *Arabidopsis thaliana* seeds was constructed and successfully screened with a portion of the spinach KAS III cDNA (16). By the same token, a 600-bp fragment of the spinach KAS III cDNA enabled the screening of a cDNA library, derived from developing embryos of *Cuphea wrigthii* (a crop accumulating mainly decanoic and dodecanoic acids), resulting in the identification of two clones, cwKAS IIIa and cwKAS IIIb (17). Then, on the basis of the sequence alignment of KAS III available, four degenerate primers were designed for polymerase chain reactions (PCR) to clone KAS III cDNAs from mRNAs of *Cuphea lanceolata* (a crop accumulating up to 87% decanoic acid) and from *Brassica napus* seeds. The resulting first strand cDNAs were used successfully as templates for the synthesis of two overlapping fragments covering the full-length ORF of the respective KAS III. The 3′-untranslated regions were determined using homologous primers in RACE-PCR reactions (Abbadi, A., and F. Spener, unpublished).

From monocotyledonous plants, a leek epidermal cDNA library has been successfully screened for KAS III cDNA with a mixture of respective cDNA clones from spinach and *Arabidopsis* (18).

Primary Structures

The *E. coli FabH* gene encodes a 317 amino acid protein with a calculated molecular mass of 33.517 Da, in good agreement with the molecular mass of 33.8 kDa determined for the native enzyme. The primary structure of FabH is 56 and 48% identical to FabH1 and FabH2 from *B. subtilis*, respectively.

The comparison of primary structures (Table 1) reveals identities between 70 and 90% within KAS III from dicots, and ~60% between dicots and monocots. Interestingly, within dicots, a stretch of four amino acids (positions 145–148, Fig. 1) is

TABLE 1
Identities of KAS III Primary Structures

Species		1	2	3	4	5	6	7	8	9
E. coli	1	—								
P. umbilicalis	2	48	—							
B. napus[a]	3	33	38	—						
C. lanceolata[a]	4	33	35	71	—					
C. wrightii a[a]	5	33	39	71	87	—				
C. wrightii b[a]	6	32	32	70	85	86	—			
S. oleracea[a]	7	33	39	79	76	71	72	—		
A. thaliana[a]	8	33	38	84	72	71	71	72	—	
A. sativum[b]	9	33	42	62	61	61	63	62	61	—

[a]Dicots.
[b]Monocot.

completely conserved; its physiologic relevance is not known. Table 1 further reveals a clear distinction between higher plant enzymes and those from *E. coli* and red alga.

All six cysteine residues in the sequence of putatively mature KAS III from plants are conserved (cysteines are shaded in Fig. 1); however, taking into account the *E. coli* and red alga enzyme, only one conserved cysteine remains. Otherwise, two arginines and a heptapeptide (see below) near the C-terminus are conserved in all bacterial and plant KAS III known to date (Fig. 1). In general terms, C-termini of KAS III proteins are more conserved than N-termini.

Tertiary Structure

The crystal structure of FabH *(E. coli* KAS III) was determined by Qiu *et al.* (19) and Davies *et al.* (20). In both cases, FabH crystallized as a homodimer and displayed a five-layered core structure, α-β-α-β-α, in which each comprises two α-helices and each β is made of a five-stranded, antiparallel β-sheet. The monomer-monomer interface within the dimer is composed of helices Nα2 and Nα3 in each monomer (Fig. 2). The overall structural organization of FabH was found to be very similar to those of FabF (KAS II) (21), *Saccharomyces cerevisiae* thiolase (22) and the *Medicago sativa* chalcone synthase (23); the monomers of these enzymes all contain the five-layered core structure in almost identical orientation, and they associate to dimers accordingly. This corresponds with their evolutionary relationship deduced from primary structures.

Sequence alignment showed that KAS III from *C. wrightii* and FabH share 37% identity. Assuming that a comparable enzymatic mechanism is operative in these two proteins, this resemblance would indicate a similar folding pattern and catalytic site architecture. On the basis of the crystallographic coordinates of *E. coli* FabH (19,20) and using the comparative protein modeling server SWISS-MODEL (24), a three-dimensional model of *C. wrightii* KAS III was generated (Fig. 2). The quality of this model has been assessed by the 3D-1D profile verification method (25) and Prosa

```
1                                                                                60
B. napus          .MANSYGFFTP SVPRSLGNKA QVPVGLSGSG FCSSLFISKR VFCSSVSESE KDAPAGNSRS
C. lanceolata     .MANALGFVGH SVP.TMGRQL NFEQMRSG.  FGSADFISKR VFCCSVVQGA GKPPRPNDRT
C. wrightii.a     .MANAYGFVGH SVP.TMKRAA QFQQMGSG.  FCSADSISKR VFCCSVVQGA DKPASGDSRT
C. wrightii.b     .MANASGFLGS SVP.ALRRAT QPQHSISSSR GSSSDFVFKR VFCCSAVQGS DRQSLGDS.
S. oleracea       .MATSYGFFSP SVPSSLNNKT SPSLGINGSG FCSHLGISKR VFCSSIEASE KHAAAGVSSS
A. thaliana       .MANASGFFTH PSIPNLRSRI HVPVRVSGSG FCVSNRFSKR VLCSSVSSVD KDASSSPS.
A. sativum       MAAASIGFTTPS ANPRIRARNF GNFGALGFLC FKERSFKRNW VGCCSVSESS SSLSYSTN.
P. umbilicalis
E. coli
```

```
                        ↓                                                       120
B. napus          ESRVSRLVSR GCKLVGCGSA VPSLQISNDD LSKFVETSDE WIATRTGIRT RRVLSGKDSL
C. lanceolata      .RSPRLVSR GCKLVGSGSA IPALQVSNDD LAKIVDTNDE WISVRTGIRN RRVLSGKDSL
C. wrightii.a     EYRTPRLVSR GCKLVGSGSA MPALQVSNDD LSKIVDTNDE WISVRTGIRN RRVLTGKESL
C. wrightii.b      .RSPRLVSR GCKLIGSGSA IPSLQISNDD LAKIVDTNDE WISVRTGIRN RRVLTGKDSL
S. oleracea       ESRVSRLVNR GCKLVGCGSA VPKLQISNDD LSKFVETSDE WIATRTGIRQ RHVLSGKDSL
A. thaliana       QYQRPRLVPS GCKLIGCGSA VPSLLISNDD LAKIVDTNDE WIATRTGIRT RRVVSAKDSL
A. sativum         .RTKRLVGM GSKLIGSGSA VPKLQIFNDD MAKIVETSDE WISVRTGIRN RRVLTGNENL
P. umbilicalis       .   ..M GVHILSTGSS VPNFSVENQQ FEDIIETSDH WISTRTGIKK SILPLLLPSL
E. coli                   .M YTKIIGTGSY LPEQVRTNAD LEKMVDTSDE WIVTRTGIRE RHIAAPNETV
```

```
                                                                                180
B. napus          TNLAAEAARN ALEMAQVNAD DIDLILMCTS TPEDLFG.SA PQIQRALGCK KNPLSYDITA
C. lanceolata     TNLATEAARK ALEMAQVDAD DVDMVLMCTS TPEDLFG.SA PQIQKALGCK KNPLSYDITA
C. wrightii.a     TNLATVAARK ALEMAQVDAN DVDMVLMCTS TPEDLFG.SA PQIQKALGCK KNPLAYDITA
C. wrightii.b     TNLASEAARK ALEMAQIDAD DVDMVLMCTS TPEDLFG.SA PQISKALGCK KNPLSYDITA
S. oleracea       VDLAAEAARN ALQMANVNPD DIDLILMCTS TPEDLFG.SA PQVQRALGCS RTPLSYDITA
A. thaliana       VGLAVEAATK ALEMAEVVPE DIDLVLMCTS TPDDLFG.AA PQIQKALGCT KNPLAYDITA
A. sativum        NGLAVEAAKG ALRMAEVEAE NVDLVIFWSS TPDDLFG.GA SRIQGDLGC. KSALAFDITA
P. umbilicalis    TKLAAEAAND ALSKASINAE DIDLIILATS TPDDLFG.SA SQLQAEIGA. TSSTAFDITA
E. coli           STMGFEAATR AIEMAGIEKD QIGLIVVATT SATHAFPSAA CQIQSMLGIK GCP.AFDVAA
```


```
1                                                                                60
B. napus          .MANSYGFFTP SVPRSLGNKA QVPVGLSGSG FCSSLFISKR VFCSSVSESE KDAPAGNSRS
C. lanceolata     .MANALGFVGH SVP.TMGRQL NFEQMRSG.  FGSADFISKR VFCCSVVQGA GKPPRPNDRT
C. wrightii.a     .MANAYGFVGH SVP.TMKRAA QFQQMGSG.  FCSADSISKR VFCCSVVQGA DKPASGDSRT
C. wrightii.b     .MANASGFLGS SVP.ALRRAT QPQHSISSSR GSSSDFVFKR VFCCSAVQGS DRQSLGDS.
S. oleracea       .MATSYGFFSP SVPSSLNNKT SPSLGINGSG FCSHLGISKR VFCSSIEASE KHAAAGVSSS
A. thaliana       .MANASGFFTH PSIPNLRSRI HVPVRVSGSG FCVSNRFSKR VLCSSVSSVD KDASSSPS.
A. sativum       MAAASIGFTTPS ANPRIRARNF GNFGALGFLC FKERSFKRNW VGCCSVSESS SSLSYSTN.
P. umbilicalis
E. coli
```


(*continued on p. 88*)

Fig. 1. Sequence alignment of KAS III. Conserved cysteines, histidine 261, and arginines 151 and 306 are shaded; the heptapeptide of the regulatory site is bold faced and shaded. Black arrow indicates the start of *C. wrightii* mature protein. *C. wrightii a,* GenBank Accession No. U15935; *C. wrightii b,* GenBank Accession No. U15934; *Spinacia oleracea,* EMBL Accession No. Z22771; *Arabidopsis thaliana,* GenBank Accession No. L31891; *Allium sativum,* GenBank Accession No. U306000; *Porphyra umbilicalis,* GenBank Accession No. 438804; *Escherichia coli,* GenBank Accession No. M77744.

```
                                                                                         300
B. napus          IEEDGLFSFD VHSDGDGRRH LNAAVKESRV DAALGSNGSV FRDFPPRQSS YSCIQMNGKF
C.lanceolata      AEEDGLFAFD LHSDGDGQKH LKSAIKEDEV DNSLGING.S IRDFPPRRSS YSCIQMNGKE
C wrightii.a      AEEDGLFAFD LHSDGDGQRH LKAAITENGI DHAVGSNG.S VSDFPPRSSS YSCIQMNGKE
C wrightii.b      AEEDGLFAFD LHSDGDGQRH LKAAIKEDEV DKALGSNG.S IRDFPPRRSS YSCIQMNGKE
S.oleracea        SEEDGMFAFD LHSDGGGGRH LNASLLNDET DAAIGNNG.A VTGFPPKRPS YSCINMNGKF
A thaliana        IEDDGLFSFD VHSDGDGRRH LNASVKESRN EGESSSNGSV FGDFPPKQSS YSCIQMNGKE
A. sativum        EDEDGLLGFD LNSDGSGQRH LNAFVSDAEH EAISNTNG   APLFPPKRST YSCIKMNGNE
P.umbilicalis     .INSILGFK LCTDGRLNSH LQLMNSPSD.   .SQQFG    LTTVPKGR.. YDSIRMNGNE
E coli            EEPGIISTH LHADGSYGEL LTLPN.ADRV N              PENS      .IHLTMAGNE

                                                                                         360
B. napus          VFRFAVRCVP QSIEAALQKA GLPSSNIDWL LLHQANQRII DAVSTRLEVP SERVISNLAN
C.lanceolata      VFRFACRCVP QSIESALGKA GLSGSNIDWL LLHQANQRII DAVATRLEVP QERIISNLAN
C.wrightii a      VFRFACRCVP QSIESALGKA GLNGSNIDWL LLHQANQRII DAVATRLEVP QERVISNLAN
C.wrightii.b      VFRFACRCVP QSIESALGKA GLNGSNIDWL LLHQANQRII DAVATRLEVP QERIISNLAN
S.oleracea        VFRFAVRCVP QSIEAALQKA GLTSSNIDWL LLHQANQRII DAVATRLEVP SERVLSNLAN
A.thaliana        VFRFAVKCVP QSIESALQKA GLPASAIDWL LLHQANQRII DSVATSLHFP PERVISNLAN
A. sativum        VFRFGWRCVP QTIQASLDDA GLSSSNIDWL LLHQANQRII DAVSTRLEIP SEKVISNLAN
P.umbilicalis     VYKFAVFQVP IVIKNCLNDV NISIDEVDWF LLHQANIRIL EAIATRLSIP LSKMITNLEN
E.coli            VFKVAVTELA HIVDKTLAAN NLDRSQLDWL VPHQANLRII SATAKKLGMS MDNVVVTLDR

                                                                   407
B.napus           YGNTSAASIP LALDEAVRSG KVKPGNTIAT SGFGAGLTWG SAIIRWG.
C.lanceolata      YGNTSAASIP LALDEAVRSG KVKPGHVIAT AGFGAGLTWG SAIVRWG.
C.wrightii.a      YGNTSAASIP LALDEAVRGG KVKAGHLIAT AGFGAGLTWG SAIVRWG.
C.wrightii.b      YGNTSAASIP LALDEAVRSG NVKPGHVIAT AGFGAGLTWG SAIIRWG.
S.oleracea        YGNTSAASIP LALDEAVRSG KVKPGNIIAT SGFGAGLTWG SSIIRWG.
A.thaliana        YGNTSAASIP LALDEAVRSG KVKPGHTIAT SGFGAGLTWG SAIVRWR.
A. sativum        YGNTSAASIP LALDEAVRNG KVKAGDTIAT AGFGAGLTWG SAIVKWG.
P.umbilicalis     YGNTSAASIP LALDEAIREK KIQPGQVVVL AGFGAGLTWG AIVLKWQ.
E.coli            HGNTSAASVP LALDEAVRDG RIKPGQLVLL EAFGGGFTWG SALVRF.
```

Fig. 1. (continued).

(26), originally built within SWISS-MODEL, as well as by Procheck (27). The results indicated that there are no unfavorable contacts between atoms in this model, and its stereochemical quality is comparable to that of the template structure of *E. coli* FabH. This is a convincing argument that a higher plant KAS III can fold like FabH.

Tissue-Specific and Development-Dependent Expression

Northern and Western blot analyses revealed tissue-specific KAS III expression in higher plants. For example, KAS III expression is elevated in spinach leaf in comparison to root (16), whereas KAS III from *C. wrightii* are expressed at higher levels in developing seed tissue than in mature seed, leaf, and root (17). In the case of monocot tissues, higher levels of expression were detected in parenchyma than in epidermis of leek. The widespread occurrence and varying expression levels indicate that this enzyme plays a role in regulating fatty acid biosynthesis, attuned to the cell and tissue of its expression (see below). This is in contrast, for example, to *Cuphea* KAS IV, which has been shown to be expressed in developing seed only when fatty acid synthesis is at its peak (11, Schütt, B.S., and F. Spener, unpublished).

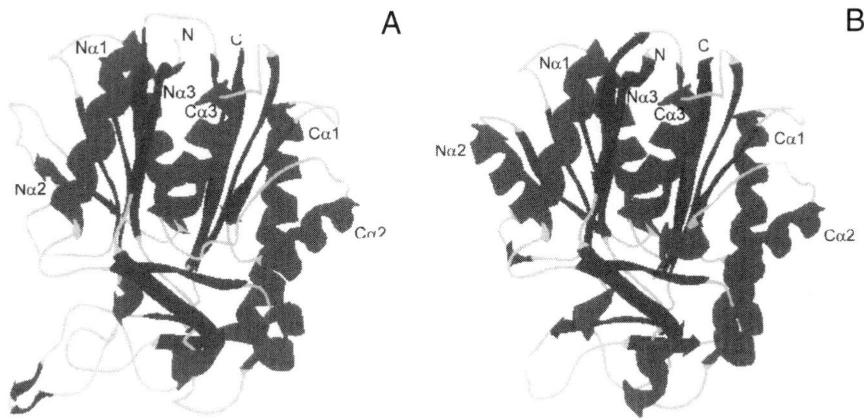

Fig. 2. Ribbon representation of the three-dimensional structure of (A) modeled *C. wrightii* KAS III monomer and of (B) *E. coli* FabH monomer. The coordinates of the template were taken from Qiu *et al.* (19). Helices are numbered as Nαx and Cαx for N- and C-termini, respectively. The figure was prepared using SWISS-Pdb Viewer 5.

Functional Characterization

Enzymatic Activities

Investigations of KAS III enzymology were conducted first with cell homogenates, then with purified native enzymes, and recently with recombinant enzymes. The consequences of the sensational discovery that KAS III use acetyl-CoA instead of acetyl-ACP for condensation with malonyl-ACP (5,6) was at first difficult to understand functionally.

It was thought earlier that, before initial condensation in fatty acid biosynthesis in bacteria and plants, a transfer of acetate from CoA to ACP, catalyzed by acetyl-CoA:ACP acetyl transferase (ACAT), was necessary to furnish acetyl-ACP for priming fatty acid synthesis. Moreover, ACAT catalysis is slow compared with that of other FAS type II enzymes and was considered to be rate limiting and a target for regulation (9). The discovery of KAS III in *E. coli* and plants then appeared to render ACAT obsolete. However, KAS III in *E. coli* (28) and plants (29,30) has ACAT activity as well, which was even more difficult to understand. Analysis of acetyl-ACP utilization by spinach chloroplasts substantiated that acetyl-ACP is only a minor intermediate in fatty acid synthesis, indicating that the role of ACAT function of KAS III is also minor; yet significant quantities of acetyl-ACP were found when spinach plants were placed in the dark (29). This accumulation can be explained by current knowledge that KAS I decarboxylates malonyl-ACP to acetyl-ACP (31,32). But if acetyl-ACP is a minor intermediate of limited importance, why has ACAT activity been maintained through evolution? It was suggested that ACAT activity plays a role in the regeneration of the acetyl-CoA pool from acetyl-ACP necessary for *de novo* fatty acid synthesis (31). This explanation can also be inferred from analysis of the substrate specificity of recombinant KAS III from *E. coli* (28) and *C. wrightii* (30); butyryl-CoA, although with lower activity than acetyl-CoA, also serves as substrate for these KAS III, which means that butyrate can be transferred from CoA to ACP for fatty acid biosynthesis.

A further difference between KAS III and other condensing enzymes is the sensitivity of the latter toward the antibiotic cerulenin. KAS I is completely inhibited by low levels (2 μmol/L) of cerulenin, KAS II and KAS IV at moderate levels (50–70 μmol/L) (3,7,8,10,11), whereas KAS III is completely insensitive to cerulenin even at high concentrations (5,8,33).

Earlier experiments with *E. coli* showed that the fungal antibiotic thiolactomycin selectively inhibited type II, dissociable, fatty acid synthases and that KAS I and II as well as ACAT were the thiolactomycin-sensitive enzymes (34). However, it is clear from what has been said above that separation of ACAT activity from KAS III activity is questionable. From the current view, the characteristics of *E. coli* ACAT, purified by Lowe and Rhodes in 1988 (35), has all the features of a KAS III; when these authors assayed this enzyme for ACAT activity, thiolactomycin did not inhibit. However, Jackowski *et al.* in 1989 (36) demonstrated that *E. coli* KAS III activity can be inhibited by thiolactomycin concentrations of ~100 μmol/L.

Application of thiolactomycin in the plant field has yielded conflicting results as well. On the one hand, the effect of thiolactomycin on purified spinach KAS III was so inconsistent that inhibition was considered to be due to an oxidation product formed from thiolactomycin (29). Yet experiments with avocado and pea KAS III revealed that thiolactomycin *per se* was inhibitory (2,37). In the work of Gulliver and Slabas (37), ACAT and KAS III activities were both inhibited by thiolactomycin, but the authors discussed the possibility that the KAS III preparations may have been contaminated by a separate ACAT. Novel, more hydrophobic analogs of thiolactomycin were shown to be more efficient inhibitors of pea KAS III than thiolactomycin (38). In the pioneering studies on enzymes of higher plant FAS type II, Shimakata and Stumpf (3,4) did not observe inhibition of ACAT by cerulenin, which agrees with the current view that this enzyme was KAS III. Taking all of these arguments together, the question whether ACAT exists in higher plants is not yet settled.

Arsenite was first reported as an inhibitor of palmitate elongation to stearate catalyzed by KAS II (39). Arsenite also inhibited purified spinach KAS III (40) as well as KAS III from *C. lanceolata* (Brück, F.M., and F. Spener, unpublished). Yet KAS III inhibition was attained with a concentration of arsenite (50 mmol/L) far higher than that used for KAS II inhibition (9).

Substrate Specificity

Although acetyl-CoA is well established as the substrate for KAS III, it remains unclear whether this enzyme can promote the elongation of short-chain acyl-ACP as well. *E. coli* KAS III was initially supposed to catalyze such elongations (5,33);

however, characterization of recombinant *E. coli* KAS III demonstrated that acetyl-CoA is the sole substrate (28). In contrast, plant KAS III indeed elongate *in vitro* short-chain acyl-ACP. On the basis of experiments with spinach leaf homogenates (29) and *C. lanceolata* seed extracts (8), the substrate specificity of KAS III appeared to range from acetyl- to butyryl-ACP. Using purified substrates and a recombinant KAS III from *C. wrightii*, we recently confirmed these observations (Fig. 3) (30). Moreover, we also showed that this KAS III could promote the elongation of propionyl- and butyryl-CoA (Fig. 3) as does recombinant *E. coli* KAS III. This offers a tool for engineering the synthesis of odd-numbered fatty acids in plants. Recently FabH1 and FabH2 from *B. subtilis* revealed a new substrate specificity because they utilize iso- and anteiso-branched acyl-CoA, in addition to acetyl-CoA, for priming fatty acid synthesis. Expression of either *B. subtilis* FabH1 and FabH2 in *E. coli* resulted in the appearance of branched-chain 17-carbon fatty acids. Thus the substrate specificity of these enzymes is an important determinant of branched-chain fatty acid biosynthesis (13).

Reaction Mechanism

KAS III operate *via* a ping-pong mechanism in which, after binding of the first substrate, a first product is released before binding of the second substrate (Fig. 4). However, the ping-pong mechanism does not indicate the order of substrate binding. In the KAS III–catalyzed reaction of *E. coli*, malonyl-ACP can bind to the free enzyme, and its decarboxylation occurs even in the absence of an acetyl-enzyme intermediate (see below). Studying the kinetics of *C. lanceolata* and *C. wrightii*

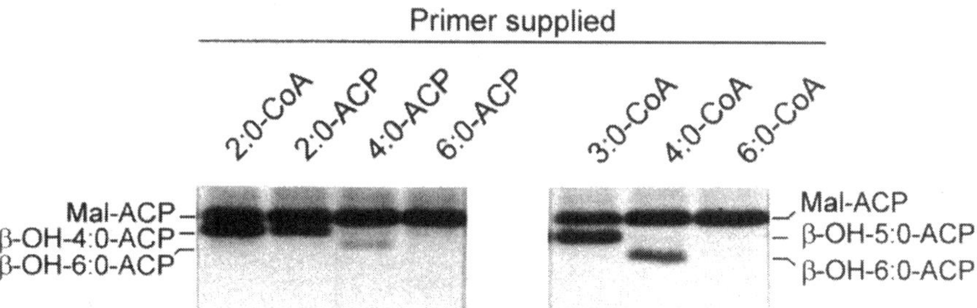

Fig. 3. Substrate specificity of recombinant KAS III from *C. wrightii*. Reaction mixtures containing 100 mmol/L sodium phosphate (pH 7.6), 20 µmol/L [2-^{14}C]malonyl-ACP, 10 µmol/L primer as indicated, 2 mmol/L NADPH, 2.5 ng KAS III and 0.6 µkat 3-β-ketoacyl-ACP reductase were incubated for 5 min at 30°C. The products of KAS III reactions, converted to β-hydroxyacyl-ACP, were separated by 2.5 mol/L urea-PAGE and visualized by autoradiography. Abbreviations: Mal-ACP, malonyl-ACP; β-OH-4:0-ACP, β-hydroxybutyryl-ACP; β-OH-5:0-ACP, β-hydroxypentanoyl-ACP; β-OH-6:0-ACP, β-hydroxyhexanoyl-ACP. From *Biochem. J. 345:*153–160 (2000) with permission of Portland Press Ltd.

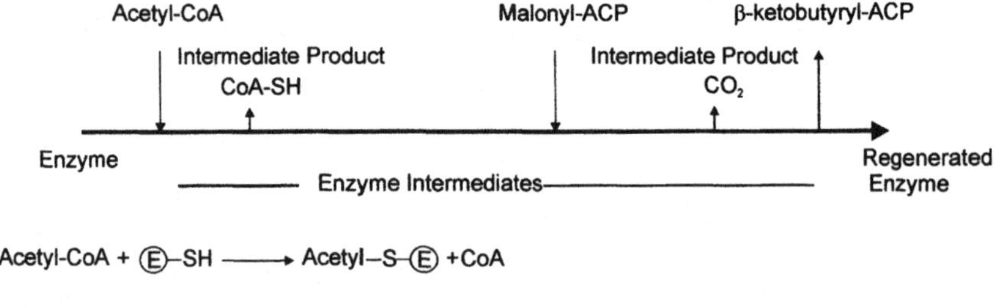

Acetyl–O–(S) + Malonyl-ACP ⟶ (E)–SH + β-Ketobutyryl-ACP + CO_2

Fig. 4. The ping-pong mechanism of KAS III reaction. The first and second half-reactions of overall *C. wrightii* KAS III catalysis are shown. Acetyl-O-E (Cys111Ser mutant) catalyzes decarboxylation of malonyl-ACP only.

KAS III–catalyzed reactions, we observed substrate binding different from that in *E. coli.* First Cys111 takes over the acetyl-group from acetyl-CoA (8,30). This was borne out by the observation that the acetylated Cys111Ser mutant of recombinant *C. wrightii* KAS III was able to decarboxylate malonyl-ACP, i.e., after formation of the acetyl-enzyme complex and release of CoA-SH. Therefore, in the first half-reaction, the formation of the acetyl-enzyme is catalyzed; in the second half-reaction, decarboxylation of malonyl-ACP and C-C bond formation by Claisen condensation occurs (30).

The catalytic residues building the active site of FabH were deduced from crystallographic structures after co-crystallization of FabH and acetyl-CoA (19) and by site-directed mutagenesis studies based on the tertiary structure (20). Three residues conserved in all KAS proteins; Cys112, His244, and Asn274 (FabH counting) lie at the bottom of a tunnel forming the active site. The structural model for *C. wrightii* KAS III reveals a similar steric conformation in the active center for the homologous residues Cys111, His261, and Asn291 (Fig. 5). His261 and Asn291 are in closest proximity to active site Cys111; the distances between the Nε of His261 and Nε of Asn291 to Sγ of Cys111 are 3.48 and 4.64 Å, respectively (Fig. 5).

For the reaction mechanism of FabH, two models have been proposed. On the basis of crystallographic data of acetyl-CoA bound to FabH, Qiu *et al.* (19) suggested that His244 plays a role as a general base in extracting a proton from Cys112, which then attacks the carboxyl-carbon of acetyl-CoA; the resulting tetrahedral intermediate is stabilized by backbone nitrogens of Gly306 and Cys112, forming the oxyanion hole. His244 may also be responsible for the binding of malonyl-ACP whose decarboxylation is aided by interaction with Asn274. On the basis of FabH crystallographic structure and subsequent experiments with enzyme mutants, Davies *et al.* (20) described a model in which both His244 and Asn274 are exclusively required for the decarboxylation of malonyl-ACP. Lowering the

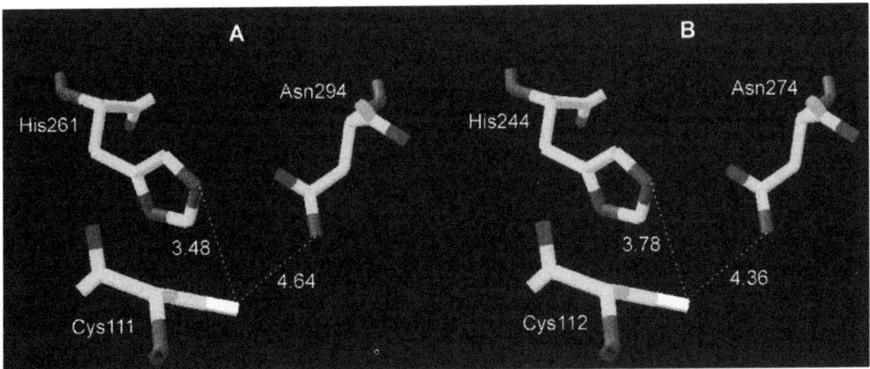

Fig. 5. Active site architecture of (A) modeled *C. wrightii* KAS III and that of (B) template *E. coli* FabH. The carbon backbone of the active site Cys and of conserved residues His and Asn are in cyan, side chain and peptide bond nitrogens in blue; sulfur and oxygen atoms are in yellow and red, respectively. Dashed lines represent the distances in Å between atoms. The figure was prepared using SWISS-Pdb Viewer 5.

pK_a of Cys112-SH to enable attack on acetyl-CoA is due to a dipole effect of the α-helix Nα3 (Fig. 2). Regardless of the approach used to define the architecture of the active site of *E. coli* FabH and the different reaction mechanisms deduced therefrom, either model considers two substrate specific binding sites and a catalytic triad to be active.

On the basis of engineered mutants of KAS III from *C. wrightii*, we recently identified Cys111 and His261 as a catalytic diad in the active site of this enzyme. We proposed a mechanistic model that ascribes to His261 general base action for acetate binding to Cys111 and general acid property for malonyl-ACP decarboxylation (for the detailed mechanism, see Fig. 6 and legend). This double function for His261 is based on kinetic evidence and the use of enzyme mutants (30). According to the structural model presented here and taking into account margins of error resulting from structural modeling, the distance of 3.48 Å between His261 and Cys111 enables this histidine to take over the catalytic double function. Asn291 is too distant (4.64 Å from Cys111) to assume one of these roles. This conclusion appears justified by our observation that the Asn291Asp (N291D) mutant of *C. wrightii* KAS III retained condensing activity (41).

Kinetics and Regulation

Multistep biosyntheses are controlled in part by a feedback mechanism, which fine-tunes the flux of metabolites through the pathway. Whenever the end product of a pathway builds up in the cell in excess of need, it inhibits primarily the enzyme of the first committed step in the pathway. Analysis of *in vivo* levels of acyl-ACP suggested earlier that, in addition to acetyl-CoA carboxylase, KAS III may play such a regulatory role (5,40,42). If KAS III is the valve that controls fatty acid biosynthesis, the key question concerns how this enzyme's activity is regulated. When KAS III in seed homogenates of *C. lanceolata* was challenged with acyl-ACP, the end product of FAS reaction, as little as 0.5 µmol/L decanoyl-ACP was able to cause 50% inhibition, for example (8). Moreover, investigations on the

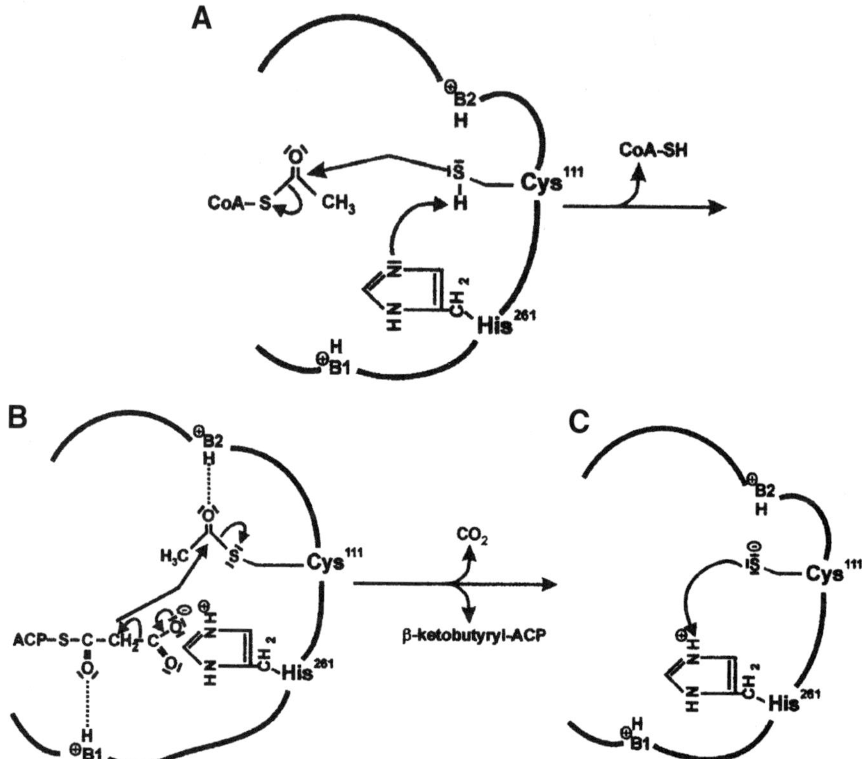

Fig. 6. Reaction mechanism of *C. wrightii* KAS III catalysis. Cys111 and His261, the catalytic diad, are part of the active site as are amino acids B1 and B2, identified as Arg151 and Arg306, that aid in acid-base catalysis according to the proposed model (30). Panel A: The thiol group of Cys111, assisted by general base catalysis of His261, attacks substrate acetyl-CoA to yield acetyl-enzyme intermediate. Panel B: The imidazolium ion of His261 generated during the formation of the acetyl-enzyme recognizes and is stabilized by the carboxylic group of malonyl-ACP; then malonyl-ACP is decarboxylated by general acid catalysis of His261, possibly enforced by B_1H^+ (the latter could also assist in positioning the carbanion). Condensing reaction occurs by nucleophilic attack of the carbanion on the enzyme-bound carboxyl-carbon of the thioester, in this case possibly aided by B_2H^+. Panel C: Upon release of the final products, the enzyme becomes regenerated by proton donation from His261 to the thiolate anion of Cys111. From *Biochem. J. 345*:153–160 (2000) with permission of Portland Press Ltd.

influence of acyl-ACP of different chain lengths on activity of recombinant KAS III from *C. wrightii* indicated that all acyl-ACP investigated inhibited KAS III, and the inhibitory potency of dodecanoyl-ACP was the highest (41). The fact that dodecanoyl-ACP was the strongest inhibitor of the C. wrightii enzyme must be seen vis-à-vis the preceding study with KAS III from *C. lanceolata*, in which decanoyl-ACP was the most efficient inhibitor (8). This fits perfectly with the fatty acid pattern found in triacylglycerols of the respective seeds, i.e., predominantly decanoic and dodecanoic acids in *C. wrightii* and decanoic acid in *C. lanceolata*. These data reinforce the concept that *Cuphea* KAS III are involved in the regulation of medium-chain fatty acid biosynthesis via a strong feedback inhibition exerted by medium-chain acyl-ACP end products of the FAS reaction in plastids of the respective seeds (8). Interestingly, the long-chain acyl-ACP end products of FAS reaction in extracts of rapeseed and spinach leaves exerted only a moderate inhibition of the respective KAS III (8). A corresponding study conducted with recombinant KAS III from *E. coli* revealed that 100 μmol/L palmitoyl- or oleoyl-ACP resulted in a 50% inhibition of the enzyme (28). Because these levels far exceeded intracellular levels observed in *E. coli*, the physiologic importance of this inhibition remains to be demonstrated.

Further kinetic analysis revealed that inhibition of KAS III from *C. lanceolata* seed extract with decanoyl-ACP (8) and recombinant *C. wrightii* KAS III with dodecanoyl-ACP (41) is uncompetitive with respect to acetyl-CoA and noncompetitive in regard to malonyl-ACP. This indicated the existence of a distinct attachment site for the inhibitor, different from the binding sites for each substrate. Taken together, there are three distinct binding sites on *Cuphea* KAS III, one each for the respective substrates and one for the regulator. In contrast, detailed kinetics of *E. coli* KAS III inhibition by long-chain acyl-ACP with respect to acetyl-CoA were mixed, i.e., a combination of competitive and noncompetitive inhibition, and with respect to malonyl-ACP, competitive (28). This indicates a total of two binding sites for substrates and regulator, one for acetyl-CoA and one for malonyl-ACP and competitive acyl-ACP.

Engineering of Mutants with New Functional Characteristics

A few plant tissues, such as palm kernels, coconut and *Cuphea* seeds, accumulate short- or medium-chain fatty acids in their storage triacylglycerols. Due to the industrial importance of oils carrying such unusual fatty acids, research in recent years has been strongly focused on the mechanisms catalyzing and regulating synthesis of medium-chain fatty acids in plants. The role of chain length–specific thioesterases (43,44) and ACP isoforms (45–47) in these mechanisms has been demonstrated. Because they catalyze the chain-lengthening reaction proper, condensing enzymes appeared to be promising candidates as well. In this context, the involvement of KAS IV, the medium-chain specific condensing enzyme, was found and characterized (8,10,11,48,49). As alluded to above, the strategic position of KAS III in initiating FAS reaction predestines this condensing enzyme to control fatty acid biosynthesis by responding allosterically to the end product of the FAS reaction. Recently, we identified this regulatory site in recombinant *C. wrightii* KAS III (41), the motif GNTSAAS near the C-terminus (Fig. 1). Mutagenesis directed to this regulatory site created mutants that were catalytically active but insensitive toward regulatory acyl-ACP. The Asn291Asp (N291D) KAS III mutant was assayed in *in vitro* experiments by employing FAS preparations from medium-chain fatty acid–producing *Cuphea lanceolata* seeds and long-chain fatty acid–producing rapeseed, respectively. Higher amounts of short-chain acyl-ACP in the case of *C. lanceolata*, and of medium-chain acyl-ACP in the case of rapeseed preparations were obtained in comparison to controls (Fig. 7 A and B, first two lanes). In *E. coli*, it is noteworthy that homologous overexpression of KAS III resulted in an accumulation of myristate and a decrease of *cis*-vaccenate (12); however, this change was later ascribed to a general metabolic phenomenon in response to cell stress (50). Overexpression of *E. coli* KAS III in transgenic rapeseed by the latter group affected constituent fatty acids of storage triglycerides in a way such that oleate decreased, whereas linoleate and α-linolenate increased. A straightforward rationale for this desaturation could not be given (50).

Our data indicate that knockout of a regulatory site in the FAS machinery of *Cuphea* could affect a rate-limiting factor and thus leads to chain shortening as a general phenomenon (Fig. 7). Changes observed in acyl-ACP amounts and composition should reflect altered regulation of fatty acid biosynthesis. The enhanced levels of butyryl-ACP produced stimulate chain lengthening up to a point at which medium-chain acyl-ACP inhibit other condensing enzymes. Thus, we consider KAS III to be responsible for maintenance of an acyl-ACP equilibrium necessary for smooth FAS reaction. Because an unregulated KAS III mutant would augment short- and medium-chain acyl-ACP in the *in vivo* pool, a carry-on effect exerted by these metabolites would inhibit condensing enzymes that accept medium-chain acyl-ACP substrates (KAS IV). Such a hypothesis was verified by the addition of decanoyl-ACP and Asn291Asp (N291D) KAS III mutant to *C. lanceolata* and *B. napus* seed extracts. In comparison to seed extracts supplemented with KAS III mutant alone (Fig. 7 A and B, middle lane), a severe inhibition of acyl-ACP chain lengthening from 6:0 to 18:0 was observed in both FAS extracts supplemented with additional decanoyl-ACP (Fig. 7 A and B, last lane). This indicates that high amounts of acyl-ACP (10 µmol/L decanoyl-ACP, in this case) upset balanced regulation of FAS machinery by inhibiting condensing enzymes that catalyze acyl-ACP elongation. On the basis of these results, we suggested that control of fatty acid synthesis by medium-chain acyl-ACP is shared by both KAS III and KAS IV in *Cuphea* (41). Whether other acyl-ACP condensing enzymes, such as KAS I or KAS II in rapeseed, for example, are subject to this type of carry-on regulation remains to be elucidated, but the example shown in Figure 7B, last lane, clearly indicates that decanoyl-ACP completely abolished medium- and long-chain acyl ACP synthesis in *B. napus* seed extracts.

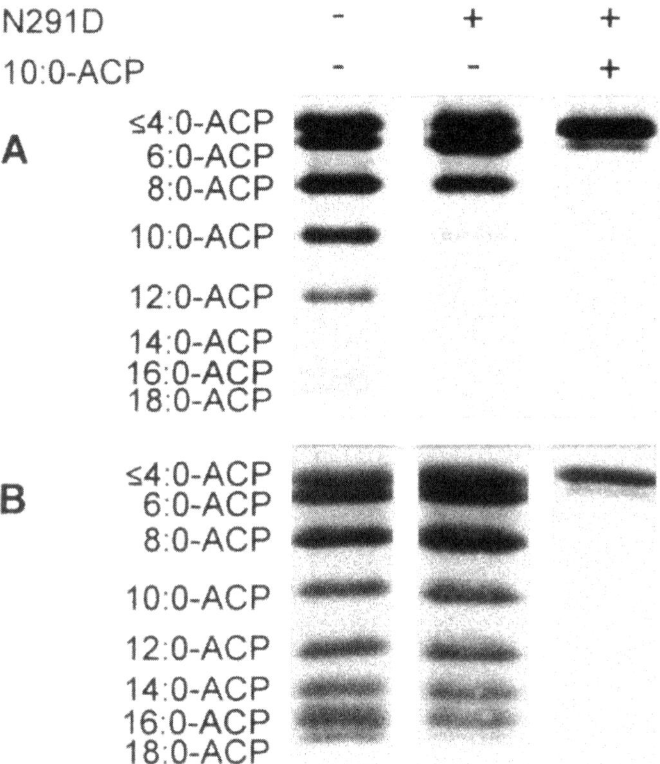

Fig. 7. Production of short- and medium-chain acyl-ACP by unregulated *C. wrightii* KAS III. FAS extracts from (A) *C. lanceolata* seeds and (B) *B. napus* were supplemented with mutant Asn291Asp (N291D) KAS III (last two lanes) and decanoyl-ACP (10 µmol/L); control was without additions (first lane). Reaction products were monitored by incorporation of [1-^{14}C]acetate from [1-^{14}C]acetyl-CoA into acyl-ACP within 20 min. From *Plant J. 24:* 1–9 (2000), with permission of Blackwell Science Ltd.

Conclusion and Perspectives

Our understanding of the regulation of fatty acid biosynthesis is much less developed than that of carbohydrate or amino acid biosynthesis pathways. We now have convincing evidence that KAS III is one key enzyme that regulates fatty acid synthesis rates, and there are indications for other control points such as KAS IV. However, this is only the beginning. With our increasing knowledge of protein chemistry, in particular the relationship between tertiary structure and enzyme activity, it is becoming possible to consider reengineering enzymes to create efficient pathways for the resynthesis of novel products. Therefore, protein engineering can help to further augment the already rich variety of fatty acid derivatives that can be produced in transgenic oilseeds. We highlighted here the creation of an enzyme with allosteric regulation knocked out. Only minimal or even single nucleotide changes are necessary in this approach. If appropriate amino acid substitutions to create desirable protein variants can be identified from *in vitro* experiments, new methodologies such as chimeric RNA/DNA oligonucleotides (51,52) have recently been developed to perform site-specific mutations of the target gene *in vivo*.

Acknowledgments

The work carried out by the authors and described here was supported by grants from the Bundesministerium für Forschung und Technologie and by the Gesellschaft zur Förderung der privaten deutschen Pflanzenzüchtung.

References

1. Ohlrogge, J.B., and J. Browse, Lipid Biosynthesis, *Plant Cell 7:* 957–970 (1995).
2. Harwood, J.L., Recent Advances in the Biosynthesis of Plant Fatty Acids, *Biochim. Biophys. Acta 31:* 7–56 (1996).
3. Shimakata, T., and P.K. Stumpf, Isolation and Function of Spinach Leaf β-Ketoacyl-(Acyl-Carrier Protein) Synthases, *Proc. Natl. Acad. Sci. USA 79:* 5808–5812 (1982).
4. Shimakata, T., and P.K. Stumpf, The Procaryotic Nature of the Fatty Acid Synthetase of Developing *Carthamus tinctorius* L. (Safflower) Seeds, *Arch. Biochem. Biophys. 217:* 144–154 (1982).
5. Jackowski, S., and C.O. Rock, Acetoacetyl-Acyl Carrier Protein Synthase, a Potential Regulator of Fatty Acid Biosynthesis in Bacteria, *J. Biol. Chem. 262:* 7927–7931 (1987).
6. Jaworski, J.G., R.C. Clough, and S.R. Barnum, A Cerulenin Insensitive Short Chain 3-Ketoacyl Acyl Carrier Protein Synthase in *Spinacia oleracea* Leaves, *Plant Physiol. 90:* 41–44 (1989).
7. Siggaard-Andersen, M., M. Wissenbach, J.A. Chuck, I. Svendsen, J.G. Olsen, and P. von Wettstein-Knowles, The fabJ-Encoded β-Ketoacyl-[Acyl Carrier Protein] Synthase IV from *Escherichia coli* Is Sensitive to Cerulenin and Specific for Short-Chain Substrates, *Proc. Natl. Acad. Sci. USA 91:* 11027–11031 (1994)
8. Brück, F.M., M. Brummel, R. Schuch, and F. Spener, *In Vitro* Evidence for Feed-Back Regulation of β-Ketoacyl Acyl Carrier Protein Synthase III in Medium-Chain Fatty Acid Biosynthesis, *Planta 198:* 271–278 (1996).
9. Shimakata, T., and P.K. Stumpf, Purification and Characterization of β-Ketoacyl-ACP Synthetase I from *Spinacia oleracea* Leaves, *Arch. Biochem. Biophys. 220:* 39–45 (1983).
10. Slabaugh, M.B., J.M. Leonard and S.J. Knapp, Condensing Enzymes from *Cuphea wrightii* Associated with Medium-Chain Fatty Acid Biosynthesis, *Plant J. 13:* 611–620 (1998).
11. Dehesh, K., P. Edwards, J. Fillatti, M. Slabaugh, and J. Byrne, KAS IV: A 3-Ketoacyl-ACP Synthase from *Cuphea* sp. is a Medium-Chain Specific Condensing Enzyme, *Plant J. 15:* 383–390 (1998).
12. Tsay, J.T., W. Oh, T.J. Larson, S. Jackowski, and C.O. Rock, Isolation and Characterisation of the β-Ketoacyl-Acyl Carrier Protein Synthase III Gene (fabH) from *Escherichia coli* K-12, *J. Biol. Chem. 267:* 6807–6814 (1992).

13. Choi, K.H., J.R. Heath, and C.O. Rock, β-Ketoacyl Acyl Carrier Protein Synthase III (FabH) Is a Determining Factor in Branched-Chain Fatty Acid Biosynthesis, *J. Bacteriol. 182:* 365–370 (2000).
14. Reith, M., A β-Ketoacyl Acyl Carrier Protein Synthase III Gene (fabH) is Encoded on the Chloroplast Genome of the Red Alga *Porphyra umbilicalis, Plant Mol. Biol. 21:* 185–189 (1993).
15. Tai, H., and J.G. Jaworski, 3-Ketoacyl Acyl Carrier Protein Synthase III from Spinach (*Spinacia oleraceae*) Is Not Similar to Other Condensing Enzymes of Fatty Acid Synthase, *Plant Physiol. 103:* 1361–1367 (1993).
16. Tai, H., D. Post-Beittenmiller, and J. G. Jaworski, Cloning of a cDNA Encoding 3-Ketoacyl-Acyl Carrier Protein Synthase III from *Arabidopsis*, *Plant Physiol. 106:* 801–802 (1994).
17. Slabaugh, M. B., H. Tai, J.G. Jaworski, and S.J. Knapp, cDNA Clones Encoding - Ketoacyl-Acyl Carrier Protein Synthase III from *Cuphea wrightii, Plant Physiol. 108:* 443–444 (1995).
18. Chen, J., and D. Post-Beittenmiller, Molecular Cloning of a cDNA Encoding 3-Ketoacyl- Acyl Carrier Protein Synthase III from Leek, *Gene 182:* 45–52 (1996).
19. Qiu, X., C.A. Janson, A.K. Konstantinidis, S. Nwagwu, C. Silverman, W.W. Smith, S. Khandekar, J. Lonsdale, and S.S. Abdel-Meguid, Crystal Structure of β-Ketoacyl-Acyl Carrier Protein Synthase III, a Key Condensing Enzyme in Bacterial Fatty Acid Biosynthesis, *J. Biol. Chem. 274:* 36465–36471(1999).
20. Davies, C., J.R. Heath, S.W. White, and C.O. Rock, A 1.8 Å Crystal Structure and Active-Site Architecture of β-Ketoacyl-Acyl Carrier Protein Synthase III (FabH) from *Escherichia coli, Structure 8:* 185–195 (2000).
21. Huang, W., J. Jia, P. Edwards, K. Dehesh, G. Schneider, and Y. Lindqvist, Crystal Structure of β-Ketoacyl-Acyl Carrier Protein Synthase II from *E. coli* Reveals the Molecular Architecture of Condensing Enzymes, *EMBO J. 17:* 1183–1191(1998).
22. Mathieu, M., Y. Modis, J.J. Zeelen, C.K. Engel, R.A. Abagyan, A. Ahlberg, B. Rasmussen, V.S. Lamzin, W.H. Kunau, and R.K. Wierenga, The 1.8 Å Crystal Structure of the Dimeric Peroxisomal 3-Ketoacyl-CoA Thiolase of *Saccharomyces cerevisiae:* Implications for Substrate Binding and Reaction Mechanism, *J. Mol. Biol. 273:* 714–728 (1997).
23. Ferrer, J. L., J.M. Jez, M.E. Bowman, R. A. Dixon, and J. P Nobel, Structure of Chalcone Synthase and the Molecular Basis of Plant Polyketide Biosynthesis, *Nat. Struct. Biol. 6:* 775–784 (1999).
24. Guex, N., and M.C. Peitsch, SWISS-MODEL and the Swiss-Pdb-Viewer: An Environment for Comparative Protein Modeling, *Electrophoresis 18:* 2714–2723 (1997).
25. Luthy, R., J. Bowie, and D. Eisenberg, Assessment of Protein Models with Three-Dimensional Profiles, *Nature 356:* 83–86 (1992).
26. Sippl, M.J., Recognition of Errors in Three Dimensional Proteins, *Proteins 17:* 355–362 (1993).
27. Laskowski, R.A., M.W. MacArthur, D.S. Moss, and J.M Thornton, PROCHECK: a Programme to Check the Stereochemical Quality of Protein Structure, *J. Appl. Crystallogr. 26:* 283–291 (1993).
28. Heath, R. J., and C.O. Rock, Inhibition of β-Ketoacyl Acyl Carrier Protein Synthase III (fabH) by Acyl-Acyl Carrier Protein in *Escherichia coli, J. Biol. Chem. 271:* 10996–11000 (1996).

29. Jaworski, J.G., H. Tai, J.B. Ohlrogge, and D. Post-Beittenmiller, The Initial Reactions of Fatty Acid Biosynthesis in Plants, *Prog. Lipid Res. 33:* 47–54 (1994).
30. Abbadi, A., M. Brummel, B.S. Schütt, M.B. Slabaugh, R. Schuch, and F. Spener, Reaction Mechanism of Recombinant 3-Oxoacyl-(Acyl Carrier Protein) Synthase III from *Cuphea wrightii* Embryo, a Fatty Acid Synthase Type II Condensing Enzyme, *Biochem. J. 345:* 153–160 (2000).
31. Heath, R. J., and C.O. Rock, Regulation of Malonyl-CoA Metabolism by Acyl-Acyl Carrier Protein and β-Ketoacyl-Acyl Carrier Protein Synthases in *Escherichia coli, J. Biol. Chem. 270:* 15531–15538 (1995).
32. Winter, E., M. Brummel, R. Schuch, and F. Spener, Decarboxylation of Malonyl-Acyl Carrier Protein by 3-Oxoacyl-(Acyl Carrier Protein) Syntheses in Plant Fatty Acid Biosynthesis, *Biochem. J. 321:* 313–318 (1997).
33. Jaworski, J.G., D. Post-Beittenmiller, and J.B. Ohlrogge, Acetyl-Acyl Carrier Protein is Not a Major Intermediate in Fatty Acid Biosynthesis in Spinach, *Eur. J. Biochem. 213:* 981–987 (1993).
34. Nishida, I., A. Kawaguchi, and M. Yamada, Effect of Thiolactomycin on the Individual Enzymes of the Fatty Acid Synthase System in *Escherichia coli, J. Biochem. 99:* 1447–1454 (1986).
35. Lowe P.N., and S. Rhodes, Purification and Characterization of [Acyl-Carrier-Protein] Acetyltransferase from *Escherichia coli, Biochem J. 250:* 789–796 (1988).
36. Jackowski, S., C.M. Murphy, J.E. Cronan, Jr., and C.O. Rock, Acetoacetyl-Acyl Carrier Protein Synthase, a Target for the Antibiotic Thiolactomycin, *J. Biol. Chem. 264:* 7624–7629 (1989).
37. Gulliver, B.S., and A.R. Slabas, Acetoacyl-Acyl Carrier Protein Synthase from Avocado: Its Purification, Characterisation and Clear Resolution from Acetyl CoA:ACP Transacylase, *Plant Mol. Biol. 25:* 179–191(1994).
38. Jones, A.L., D. Herbert, A.J. Rutter, J.E. Dancer, and J.L. Harwood, Novel Inhibitors of the Condensing Enzymes of the Type II Fatty Acid Synthase of Pea (*Pisum sativum*), *Biochem. J. 374:* 205–209 (2000).
39. Harwood, J.L., and P.K. Stumpf, Fat Metabolism in Higher Plants. 43. Control of Fatty Acid Synthesis in Germinating Seeds, *Arch. Biochem. Biophys. 142:* 281–291 (1971).
40. Clough, R.C., A.L. Matthis, S.R. Barnum, and J.G. Jaworski, Purification and Characterization of 3-Ketoacyl-Acyl Carrier Protein Synthase III from Spinach. A Condensing Enzyme Utilizing Acetyl-Coenzyme A to Initiate Fatty Acid Synthesis, *J. Biol. Chem. 267:* 20992–20998 (1992).
41. Abbadi, A., M. Brummel, and F. Spener, Knockout of Regulatory Site of 3-Ketoacyl-ACP Synthase III Enhances Short- and Medium-Chain Acyl-ACP Synthesis, *Plant J. 24:* 1–9 (2000).
42. Post-Beittenmiller, D., J.G. Jaworski, and J.B. Ohlrogge, *In Vivo* Pools of Free and Acylated Acyl Carrier Proteins in Spinach—Evidence for Sites of Regulation of Fatty Acid Biosynthesis, *J. Biol. Chem. 266:* 1858–1865 (1991).
43. Voelker, T., A. Worrell, L. Anderson, J. Bleibaum, C. Fan, D. Hawkins, S. Radke, and H. Davies, Fatty Acid Biosynthesis Redirected to Medium-Chain in Transgenic Oilseed Plants, *Science 257:* 72–74 (1992).
44. Töpfer, R., N. Martini, and J. Schell, Modification of Plant Lipid Synthesis, *Science 268:* 681–686 (1995).

45. Shintani, D.K., and J.B. Ohlrogge, The Characterization of a Mitochondrial Acyl Carrier Protein Isoform Isolated from *Arabidopsis thaliana, Plant Physiol. 104:* 1221–1229 (1994).
46. Schütt, B.S., M. Brummel, R. Schuch, and F. Spener, The Role of Acyl Carrier Protein Isoforms from *Cuphea lanceolata* in the *De-Novo* Biosynthesis of Medium-Chain Fatty Acids, *Planta 205:* 263–268 (1998).
47. Suh, M. C, D.J. Shultz, and J.B. Ohlrogge, Isoforms of Acyl Carrier Protein Involved in Seed-Specific Fatty Acid Synthesis, *Plant J. 17:* 679–688 (1999).
48. Fuhrmann, J., and K.P. Heise, Factors Controlling Medium-Chain Fatty Acid Synthesis in Plastids from Maturing *Cuphea* Embryos, *Z. Naturforsch. 48c:* 616–622 (1993).
49. Schuch, R., M. Brummel, and F. Spener, β-Ketoacyl Acyl Carrier Protein (ACP) Synthase III in *Cuphea lanceolata* Seeds: Identification and Analysis of Reaction Products, *J. Plant Physiol. 143:* 556–560 (1994).
50. Verwoert , I.I.G.S., K.H. van der Linden, M.C. Walsh, H.J.J. Nijkamp, and A.R. Stuitje Modification of *Brassica napus* Seed Oil by Expression of the *Escherichia coli fabH* Gene, Encoding 3-Ketoacyl-Acyl Carrier Protein Synthase III, *Plant Mol. Biol. 27:* 875–886 (1995).
51. Beetham, P.R., P.B. Kipp, X. L. Sawycky, C.J. Arntzen, and G.D. May, A Tool for Functional Plant Genomics: Chimeric RNA/DNA Oligonucleotides Cause *In Vivo* Gene-Specific Mutations, *Proc. Natl. Acad. Sci. USA 96:* 8774–8778 (1999).
52. Zhu, T., D.J. Peterson, L. Tagliani, G. St. Clair, C.L. Baszczynski, and B. Bowen, Targeted Manipulation of Maize Genes *In Vivo* Using Chimeric RNA/DNA Oligonucleotides, *Proc. Natl. Acad. Sci. USA 96:* 8768–8773 (1999).

Chapter 11

The Rice Genome Project in Japan

Takuji Sasaki

Rice Genome Research Program, National Institute of Agrobiological Resources, Ibaraki 305-8602, Japan

Introduction

Rice is one of the major cereal crops, providing food for about half of the world's population (1). The global population reached 6 billion in 1999 and will continue to increase by 1 billion every 12 years. The growing demand for rice, especially in areas where it is used as a staple food, i.e., Asia, Africa, and Latin America (2), dictates the need for scientists to create innovative breeding concepts. The first goal is to ensure a secure and sustainable food supply by increasing rice productivity even under unfavorable cultivation conditions such as drought, salinity, or flood. Rice breeding has a long history that varies according to prevailing local traditions. In the absence of knowledge of the laws of inheritance, farmers have relied on practical breeding practices that have produced a considerable number of rice varieties worldwide. The second concept is that breeding is based on the laws of inheritance in combination with empirical knowledge. This approach is evident in almost all breeding strategies employed by any government or private entity. However, almost all of the important agronomic traits in rice are controlled not by a single gene but by multiple, complex genes. Collectively known as quantitative trait loci (QTL), these traits are difficult to analyze genetically and to modify by conventional breeding technology. To overcome this difficulty, the third concept in breeding requires the use of genetic material (DNA), which carries all of the information on inheritance for direct or indirect genetic analysis and breeding.

The Japanese government launched an extensive Rice Genome Project (RGP) in 1991 to obtain complete information on inheritance in rice plants; a team of scientists was organized to perform the task at one place in Tsukuba City, Japan (3). From its start until 1998, RGP focused on establishing a fine molecular genetic map, then converting this genetic map to a physical map, and collecting enormous amounts of expressed sequence tags (EST). The molecular data obtained by these tools were applied, for example, to map-based cloning of several genes corresponding to phenotypes caused by a single Mendelian factor or to advanced backcrossing strategies for detailed dissection of QTL with DNA markers. The most beneficial use of these combined tools was the development of a reliable physical map for genome sequencing. Construction of a sequence-ready physical map and the genome sequencing stage started in 1998 as integral parts of the second phase of RGP (4). In addition to genome sequencing, a program was launched for the identification of rice gene function by forward and/or reverse genetic methods, for monitoring gene expression by microarray of rice cDNAs, and for understanding the translation products as proteomes to offer a broader application of rice genome analysis. In this article, I will describe our major research findings on the molecular genetic analysis of rice plants during the past 10 years and their expected applications and effects in the future.

Dissecting the Rice Genome on the Basis of Genetic Events

Before the development of the concept of polymorphism in a DNA sequence as detected by restriction enzymes [referred to as restriction fragment length polymorphism (RFLP)], allelic phenotypes segregated by the Mendelian law of inheritance were chosen as markers to calculate genetic distances on the basis of recombination frequencies. A conventional genetic map of rice was constructed by accumulating such data with the most plausible ordering and distances between pairs of phenotypes. However, each datum was obtained from different cultivars and populations so that accuracy varied accordingly. This resulted in only a rough linkage map, indicating the relative position of ~200 traits within 12 linkage groups (5).

RFLP was first applied for accurate tagging of the famous and tragic human heritable Huntington's chorea. This success proved that RFLP was a reliable and a reproducible tool for establishing a molecular genetic map of any species (6). In the case of rice plants, several types of segregated rice populations and probes to detect RFLP have been used. RGP used an F2 population composed of 186 progenies from a *japonica* variety Nipponbare and an *indica* variety Kasalath to construct a genetic map with 2275 markers (7). This map is the finest rice genetic map available to date, and ~70% of the markers are derived from rice expressed genes (cDNAs). The markers on this map are very reliable because their locations are based on genetic distance. Without this map, genetic identification and localization of unassigned genes and traits on the rice genome would not be possible.

The information derived from the genetic map has various applications as follows: (i) the markers can be used as probes to pick up DNA fragments by hybridization or polymerase chain reaction (PCR) for constructing a rice physical map; (ii) genetic mapping of rice markers to other cereal crops and *vice versa* helps to clarify the common nature of gene ordering, or synteny among Gramineae (8); (iii) fine resolution of recombination events facilitates detection of the candidate genome region of a gene corresponding to a trait, and to further identify the gene (9); (iv) the combined use of an advanced backcrossing method using several markers allows the isolation of each component contributing to quantitative trait loci (QTL) and the final identification of each gene involved in the QTL (10).

We challenged the relevance of these applications and obtained several [illegible]nifi- cant results. As for synteny, we collaborated with researchers working o[illegible] foxtail millet, barley, or corn and clarified the existence of conserved ge[illegible] ing within blocks comprising each genome (11). This finding establishe[illegible] model system for genome analysis, particularly for genome sequencing because it has the smallest genome size among the cereal crops. As for gene identification, we successfully cloned *Xa1* (12), the resistance gene to bacterial blight, *Pib* (13), the resistance gene to rice blast, and *d1* (14), gibberellin-insensitive dwarfism. These genes could not be identified without the development of molecular markers. These newly discovered genes should be utilized further in the study of physiology and biochemistry of signal transduction pathways and/or interactive modes among participating components on disease resistance or hormonal regulation mechanisms in rice plants. A detailed analysis of QTL has focused on the flowering time of rice plants (15). This characteristic is very important for adapting a rice variety to conditions such as the length of the local day; it is difficult to manipulate, however, in part because of to the complex interaction of other QTL such as photoperiod sensitivity and vegetative growth rate. We have identified at least 13 loci for QTL of flowering time in a population obtained by crossing a *japonica* and an *indica* variety (M. Yano, personal communication). These loci are isolated as a single Mendelian factor by an advanced backcrossing method; most of them have been identified as having a corresponding gene. These findings are fundamental knowledge for managing flowering time in rice plants and illustrate the validity of DNA markers to analyze a complex genetic event.

Dissecting the Rice Genome on the Basis of Molecular Markers

A genetic map is constructed on the basis of distances calculated by recombination frequency between two loci on the genome. This does not give the actual physical distance, which is measured by the number of nucleotides between two loci. Converting the genetic map to a physical map is indispensable for obtaining useful and direct information, as well as materials on molecular breeding and the biology of rice. The physical map is constructed with pieces of genomic DNA fragments similar to a jigsaw puzzle. To realize the molecular architecture of the genome, a library of genomic DNA fragments with an average size of several hundred kilobase pairs is initially constructed using an artificial chromosome propagated in yeast (YAC) or bacteria (BAC or PAC). YAC are suitable for carrying larger-sized fragments (300–500 kb) but frequently generate chimera of these fragments in a clone. On the other hand, BAC/PAC can carry smaller-sized fragments (100–200 kb) but can stably propagate the inserted fragments. We used YAC to construct a basic physical map and PAC/BAC for a map to promote genome sequencing. Our YAC library is composed of ~7000 clones with an average insert size of 350 kb DNA, derived from Nipponbare, which is one of the parents of the F2 population for linkage analysis (16). These clones were spotted onto a nylon membrane and probed by DNA markers on the genetic map. The DNAs from positive YAC were restricted by the same enzyme and probed by the same probe that was used to genetically map the corresponding RFLP. Only the YAC giving the corresponding Southern pattern of Nipponbare were aligned correctly along the linkage map. Our

YAC physical map was constructed by using almost all of the 2275 RFLP markers on our linkage map and covered 63% of the rice genome (17). The remaining 37% of the rice genome must be filled by using more markers or by walking using the end portion of a YAC flanking on a gap.

We also generated many ESTs, which were derived from partial sequences of randomly cloned rice cDNA clones (18). These ESTs were used to upgrade the YAC physical map by PCR screening of YAC using the 3′-UTR of each EST as a specific primer (18). The selected YACs were aligned on the skeleton YAC physical map described above using the SEGMAP software. This new mapping strategy has created ~6000 new PCR markers of expressed genes and increased the coverage of the rice genome with YAC from 63 to >80% (J. Wu, personal communication). Although this mapping method cannot define the order of mapped EST on each harboring YAC, ordering information of overlapping YAC containing RFLP markers may be known in some cases. Even if the ordering is ambiguous, EST mapping enhances gene identification by map-based cloning or by gene disruption and, more importantly, is indispensable for constructing a new and accurate sequence-ready physical map.

Dissecting the Rice Genome on the Basis of Nucleotide Sequence

To construct a sequence-ready physical map, we prepared another type of library of rice genomic DNA because YAC contains 16 yeast chromosomes in addition to an alien rice DNA fragment. In some cases, this alien fragment is a chimera of more than two different fragments. The newly chosen vector is a P1-derived artificial chromosome (PAC), which can harbor a DNA fragment of 50–300 kb without generating a chimera. Our PAC library is composed of 71,000 clones with an average insert size of 112 kb of Nipponbare DNA restricted by *Sau*3A1 (19).

For constructing a sequence-ready physical map, we employed a marker-based PCR screening method utilizing EST and RFLP on the map as resources of primer information. The advantages of this method are as follows: (i) the assignment of selected PAC on the corresponding position on the rice genome is accurate; and (ii) the genome regions covered by these PAC are also expected to cover regions with a high density of expressed genes. The latter characteristic is favorable for sequencing gene-rich regions in the rice genome. The disadvantages are that genomic regions without any markers cannot be screened for locating corresponding PACs and that additional information such as fingerprints or end-sequences of these PAC is required to ascertain the overlap and order of selected PAC.

PAC clones selected by PCR primers and checked thereafter for their minimum mutual overlapping by fingerprinting are subjected to sequencing by a shotgun method. Shotgun libraries are usually composed of 2000 subclones, each of 2- and 5-kb fragment size, using pUC18 as a plasmid vector. The isolated plasmids from these clones are subjected to a cycle sequencing reaction by a Dye-terminator method, and the reaction products are analyzed for their sequences from both ends by an automated sequencer such as the Applied Biosystems model ABI3700. The 3000–4000 fragment sequences obtained are assembled using the Phred/Phrap software. The completed sequences are annotated by a combination of gene prediction and similarity search tools, such as Genscan and BLAST.

The sequencing of the rice genome is a combined effort of an international collaborative organization, IRGSP (International Rice Genome Sequencing Project) (20). This consortium has established policies and guidelines, such as how to sequence the rice genome or how to publish the sequence data. To date, 11 countries participate in IRGSP and share the sequencing of the 12 rice chromosomes. For example, Japan is in charge of sequencing chromosomes 1, 6, 7, 8, and the United States is in charge of chromosomes 3, 10, and 11. The updated progress of sequencing at RGP and IRGSP is available at the http://rgp.dna.affrc.go.jp/GenomeSeq.html and http://rgp.dna.affrc.go.jp/Seqcollab.html websites, respectively. A newly developed database INE (INtegrated rice genome Explorer) with an integrated map view has been especially designed to show fully annotated genome sequences and other relevant information (21).

Very recently, Monsanto, a private company, decided to offer their data, which comprise a sequence-ready BAC (bacterial artificial chromosome) physical map covering 85% of the rice genome and the rough draft sequence of each BAC on this map. Because the BAC physical map is constructed mainly with fingerprints and sequence-tag-connectors (STC) derived from BAC-end sequence information, these data are expected to facilitate gap-filling on our PAC physical map. Both types of sequence-ready physical maps must be complementary. The rough draft sequences will also help to analyze the sequences of corresponding BAC in an accurate manner. IRGSP aims to complete the sequencing of the entire rice genome by 2004.

Dissecting the Rice Genome as a Piece of a Functional Unit

The sequence information alone is insufficient in understanding the meaning of what is written in the genetic codes. Although several computer software programs can predict the position of open reading frames or cis-elements controlling gene expression, the information must be verified by experimental evidence. Strategies such as forward or reverse genetics must be developed to identify expressed genes.

In forward genetics, a rice population segregating with a single Mendelian factor is required for a target phenotype. The quality of this analysis absolutely defines whether the trial is successful or not, even when using several DNA markers on a fine genetic map. Once the trait is accurately tagged with DNA markers, the gene isolation procedure will utilize the physical map and the sequence information for final identification of the candidate gene as described in the previous section. This strategy is also applicable to identify genes involved in a trait caused by multigene effects or by a quantitative trait. In this case, there exists no mutant trait, and only the difference of gene effects between rice varieties is observed.

Many agronomically important traits are classified under this category such as heading date (flowering time), grain yield, or culm height. We have targeted genes responsible for photoperiod sensitivity in heading date and have succeeded in isolating two of the eight genes involved in this characteristic.

Reverse genetics is applicable for detecting artificial gene-disrupted mutant by a sequence-known fragment. In some rice varieties, an endogenous retrotransposon, *Tos17*, is activated during cell culture and is steadily inserted in several positions in the genome after regeneration (22). The key feature of this method is that the mutated rice plants are not transgenic. This means that the regenerated plants can be cultivated in an open paddy field. Before the discovery of the rice retrotransposon, other transposable elements such as *Ac/Ds* corn were often used (23). In this case, however, the resulting rice progenies were transgenic plants and wide areas of regulated paddy fields were essential for their propagation. The disrupted genes can be identified when flanking sequences are compared to *Tos17* with our EST. If this tagged line indicates a mutated phenotype, we can then define the function of the gene. The mutants produced by *Tos17* are also used for identification of genes by forward genetics as described above. To date, a gene corresponding to a viviparous or brittle culm has been identified (H. Hirochika, personal communication).

These two strategies are now principally employed in Japan for the elucidation of gene function. Other types of tools are still valuable to disrupt genes such as the use of chemicals that induce mutation or irradiation. The number of alleles should be increased for each gene for better understanding of genetic and physiologic functions. In this regard, tools to produce mutants are indispensable.

Once genes corresponding to traits are identified, the next step is to clarify the controlling unit responsible for expression of these genes. Genome sequence information can facilitate this process because we can speculate on a sequence empirically and manipulate it to finally identify the responsible unit. The accumulation of an enormous amount of information through many experiments requires improved knowledge of data processing and interpretation through a computationally predicted information system. Although it is a tedious process to achieve reliable information, genome analysis must be accomplished thoroughly and carefully to achieve its optimum application not only for rice breeding but also for breeding of other cereal crops.

Application of Genome Sequencing to Agricultural Sciences

There are numerous applications of the results obtained from each step of genomic dissection to agricultural sciences. First is the application of DNA markers for selection of favorable traits, also referred to as marker-assisted selection (MAS). A trait controlled by a single gene does not necessarily require this method for critical selection because the gene effect is straightforward and plants can be easily selected by their phenotypes. However, if multigenes such as QTL govern a phenotype, selection only by phenotype may not be sufficient. Only MAS can specifically discriminate plant individuals carrying the genes in a desirable combination. To perform MAS successfully, an accurate genetic analysis is required to tag each QTL component. Many DNA markers generated at RGP by rice genome analysis must be used. In addition, even for single-gene phenotypes, MAS is also effective for judging the accumulation of genes after ordinary crossing. This approach is called gene pyramiding and is often used in disease resistance genes (24). The second application of genome analysis is to propagate modified plants by transformation. A Japanese study demonstrated that *Agrobacterium* can transfer genes in rice (25). Thereafter, this system has been useful to generate many types of transformants for experimental purposes. The final and possibly the most significant application of rice genome analysis is its contribution of vast amounts of knowledge to the basic sciences such as genetics, biology, and to the newly emerging field of bioinformatics. We are now in a chaotic situation to extract valuable information and translate it into a whole new framework for biological investigations or discoveries, as well as interpretations to generate new paradigms in the near future; we are at a turning point in genome research.

References

1. White, P.T., Rice: The Essential Harvest, *Natl. Geogr. Mag. 185:* 48–79 (1994).
2. http://www.riceweb.org/Ricetabindex.htm.
3. Rice Genome, vol.1, http://rgp.dna.affrc.go.jp/rgp/ricegenomenewslet/nl1.html, 1992.
4. Sasaki, T., Rice Genome Research in Japan as a Frontrunner in Crop Genomics, *Sci. Technol. Jpn. 18:* 18–21 (1999).
5. Kinoshita, T., Report on Gene Symbolization, *Rice Genet. Newslett. 12:* 9–153 (1995).
6. Botstein, D., R.L. White, M. Skolnick, and R.W. Davis, Construction of a Genetic Linkage Map in Man Using Restriction Fragment Length Polymorphisms, *Am. J. Hum. Genet. 32:* 314–331 (1980).
7. Harushima, Y., M.Yano, A. Shomura, M. Sato, T. Shimano, Y. Kuboki, T. Yamamoto, S.Y. Lin, B.A. Antonio, A. Parco, H. Kajiya, N. Huang, K. Yamamoto, Y. Nagamura, N. Kurata, G.S. Khush, and T. Sasaki, A High-Density Rice Genetic Linkage Map with 2275 Markers Using a Single F2 Population, *Genetics 148:* 1–16 (1998).
8. Moore, G., K.M. Devos, Z. Wang, and M.D. Gale, Cereal Genome Evolution, Grasses, Line Up and Form a Circle, *Curr. Biol. 5:* 737–739 (1995).
9. Tanksley, S.D., M.W. Ganal, and G.B. Martin, Chromosomal Landing: A Paradigm for Map-Based Cloning in Plants with Large Genomes, *Trends Genet. 11:* 63–68 (1995).
10. Yano, M., and T. Sasaki, Genetic and Molecular Dissection of Quantitative Traits in Rice, *Plant Mol. Biol. 35:* 145–153 (1997).
11. Devos, K.M., and M.D. Gale, Comparative Genetics in the Grasses, *Plant Mol. Biol. 35:* 3–15 (1997).
12. Yoshimura, S., U. Yamanouchi, Y. Katayose, S. Toki, Z.X. Wang, I. Kono, N. Kurata, M. Yano, N. Iwata, and T. Sasaki, Expression of Xa1, a Bacterial Blight-Resistance Gene in Rice, Is Induced by Bacterial Inoculation, *Proc. Natl. Acad. Sci. USA 95:* 1663–1668 (1998).
13. Wang, Z.X., M. Yano, U. Yamanouchi, M. Iwamoto, L. Monna, H. Hayasaka, Y. Katayose, and T. Sasaki, The Pib Gene for Rice Blast Resistance Belongs to the

Nucleotide Binding and Leucine-Rich Repeat Class of Plant Disease Resistance Genes, *Plant J. 19:* 55–64 (1999).

14. Ashikari, M., J. Wu, M. Yano, T. Sasaki, and A. Yoshimura, Rice Gibberellin-Insensitive Dwarf Mutant Gene Dwarf 1 Encodes the Alpha-Subunit of GTP-Binding Protein, *Proc. Natl. Acad. Sci. USA 96:* 10284–10289 (1999).
15. Yamamoto, T., H. Lin, T. Sasaki, and M.Yano, Identification of Heading Date Quantitative Trait Locus Hd6 and Characterization of Its Epistatic Interaction with Hd2 in Rice Using Advanced Backcross Progeny, *Genetics 154:* 885–891 (2000).
16. Umehara,Y., A. Inagaki, H. Tanoue, Y. Yasukochi, Y. Nagamura, S. Saji, Y. Otsuki, T. Fujimura, N. Kurata, and Y. Minibe, Construction and Characterization of a Rice YAC Library for Physical Mapping, *Mol. Breed. 1:* 79–89 (1995).
17. Saji, S., Y. Umehara, B.A. Antonio, H. Yamane, H. Tanoue, T. Baba, H. Aoki, N. Ishige, J. Wu, K. Koike, T. Matsumoto, and T. Sasaki, A Physical Map with Yeast Artificial Chromosome (YAC) Clones Covering 63 Percent of the 12 Rice Chromosomes, *Genome* Feb. (2001).
18. Sasaki, T., J. Song, Y. Koga-Ban, E. Matsui, F. Fang, H. Higo, H. Nagasaki, M. Hori, M. Miya, E. Murayama-Kayano, T. Takiguchi, A. Takasuga, T. Niki, K. Ishimaru, H. Ikeda, Y. Yamamoto, Y. Mukai, I. Ohta, N. Miyadera, I. Havukkala, and Y. Minobe, Toward Cataloguing All Rice Genes: Large-Scale Sequencing of Randomly Chosen Rice cDNAs from a Callus cDNA Library, *Plant J. 6:* 615–624 (1994).
19. Baba,T., S. Katagiri, H. Tanoue, R. Tanaka,Y. Chiden, S. Saji, M. Hamada, M. Nakashima, M. Okamoto, M. Hayashi, S. Yosiki, W. Karasawa, M. Honda, Y. Ichikawa, K. Aita, M. Ikeno, T. Ohta, Y. Umehara, T. Matsumoto, P.J. de Jong, and T. Sasaki, Construction and Characterization of Rice Genomic Librares: PAC Library of *Japonica* Variety, Nipponbare and BAC Library of *Indica* Variety, Kasalath *Bull., Natl. Inst. Agrobiol. Resour. 14:* 41–49 (2000).
20. Sasaki, T., and B. Burr, International Rice Genome Sequencing Project: The Effort to Completely Sequence the Rice Genome, *Curr. Opin. Plant Biol. 3:* 138–141 (2000).
21. Sakata, K., B.A. Antonio, Y. Mukai, H. Nagasaki, Y. Sakai, K. Makino, and T. Sasaki, INE: A Rice Genome Database with an Integrated Map View, *Nucleic Acids Res. 28:* 97–101 (2000).
22. Hirochika, H., Retrotransposons of Rice as a Tool for Forward and Reverse Genetics, in *Molecular Biology of Rice,* edited by K. Shimamoto, Springer Verlag, Berlin, 1999, pp. 43–58.
23. Izawa, T. & Shimamoto, K. (1999) Ac/Ds as Tools for Rice Molecular Biology, in *Molecular Biology of Rice,* edited by K. Shimamoto, Springer Verlag, Berlin, 1999, pp. 59–76.
24. Huang, N., E.R. Angeles, J. Domingo, S. Magpanthy, S. Singh, G. Zhang, N. Kumaravadivel, J. Bennet, and G.S. Khush, Pyramiding of Bacterial Blight Resistance Genes in Rice: Marker-Assisted Selection Using RFLP and PCR, *Theor. Appl. Genet. 95:* 313–320 (1997).
25. Hiei, Y., T. Komari, and T. Kubo, Agrobacterium-Mediated Transformation, in *Molecular Biology of Rice,* edited by K. Shimamoto, Springer Verlag, Berlin, 1999, pp. 235–256.

Chapter 12

Development of a Genetically Based Seed Technology Protection System

Melvin J. Oliver and Jeff Velten

USDA-ARS, Plant Stress and Water Conservation Laboratory, Lubbock, TX 79415

Introduction

The use of plant biotechnology for crop improvement has been hampered by three concerns: (i) the need to prevent the escape of transgenes into wild species; (ii) the need to protect the introduced technology from unauthorized use; and (iii) the inability to recuperate the investment in technology development. We have developed a genetic system that addresses these concerns. The system, designated the Technology Protection System (TPS), employs three elements, a chemically inducible promoter or derepressible promoter system, a site-specific recombinase, and a late embryogenesis–specific promoter. A chemical seed treatment activates the site-specific recombinase gene during germination; the site-specific recombinase is synthesized and removes a blocking DNA sequence situated between the late embryogenesis promoter and a sequence that encodes a germination disruption protein. This results in the formation of an active gene that directs the synthesis of a germination disruption protein only in the new embryos that develop on the plant and only during the latter stages of seed maturation (after the seed has fully reached its commercial potential). Therefore, in a plant in which the genetic technology protection system is active, commercially useful seeds are produced. However, these seeds are incapable of germination. In practical terms, this means that the chemically treated seeds themselves can be germinated and grown but, due to their production of nongerminable descendant seed, they cannot form the basis of a new crop. In addition, the production of nongerminable descendant seed prevents the spread of genetic traits *via* seed escapes or seed carry-over during crop rotations. Pollen from TPS-activated plants also carries the dominant germination defective trait; thus, linked traits carried by the active TPS-containing plant cannot be transmitted to closely related weedy species via hybridization (any seed produced by the wild species through fertilization by TPS pollen is incapable of germination).

TPS for Self-Pollinating Crops

The particular embodiment of the Technology Protection System that is under development and being tested currently was designed specifically for self-pollinating crops, in particular cotton, and is described in detail in Oliver *et al.* (1), U.S. Patent 5,723,765.

The TPS genetic constructs are, by necessity, complex and are manufactured as two components, each of which is introduced into separate transgenic founder lines. Crossing the two founder lines generates plants containing the TPS in an inactive state. This cross is made only in the establishment phase of TPS; the seed company or the producer is not required to cross plants to obtain TPS. As shown in Figure 1 the constructs that form the basis of the two components currently being tested are as follows: (i) Late Embryogenesis Protein (LEA) promoter-LOX-tetracycline (tet) repressor gene-LOX-protein synthesis inhibitor protein coding sequence/PolyA site; and (ii) 35S promoter containing tet operator sites-CRE site-specific recombinase.

Within these two types of constructs, we are testing two different LEA promoters (see Ref. 1, U.S. Patent 5,723,765) and two different protein synthesis inhibitor proteins, i.e., a ribosomal inhibitory protein (RIP) called saporin (2) and a ribonuclease from *Bacillus amyloliquefaciens* (Barnase). This embodiment of TPS uses a repressible promoter system, the tet repressor system first described to be functional in plants by Gatz and Quail (3), to act as the trigger for DNA excision *via* the site-specific recombinase CRE, a bacteriophage-derived enzyme that can also operate in plant cells [first demonstrated by Dale and Ow (4) and Odell *et al.* (5)].

Once the complete TPS is in a single plant, the constructs are basically inactive and the plant is unaffected by their presence in the genome (Fig. 1). Upon activation of the TPS by the treatment of seed with low levels of tetracycline, the tet repressor gene is removed by CRE and the LEA promoter is joined to the germination disrupter sequence (Fig. 1). The seed is cleaned of excess tetracycline and dried, ready for sale. The seed will germinate and produce normal plants that will allow for the full potential of the crop to be realized. However, at the late stage of seed development, the LEA promoter is activated and the seed cells produce the protein synthesis inhibitor protein. In effect, this makes the seeds incapable of germination (because this event takes place very late in development, the seed will appear normal and will be unaffected in a commercial sense). Although the seeds produced by an activated TPS plant would be fully developed, for sale as feed or grain, they would not germinate. In addition, because the activation of the TPS occurs during inhibition of the progenitor seed and is targeted to the cells of the embryonic meristem, the pollen from an activated TPS plant will contain the activated LEA-protein synthesis inhibitor protein gene. This means that if this pollen fertilizes an ovule on a neighboring plant (be it a crop or weed species), the seed it produces will be nonviable. In effect, this makes an activated TPS plant an evolutionary dead end (both seed and pollen are effectively nonviable), incapable of spreading transgenes into the environment (Fig. 2). The above system is only one of many possible embodiments of TPS and is one we are currently testing. Additional embodiments will consist of alternative germination disruption genes or other induction systems, for example.

To date, we have generated tobacco plants that contain the complete TPS system (Fig. 1). Seeds from these plants have been treated with tetracycline, and various parts

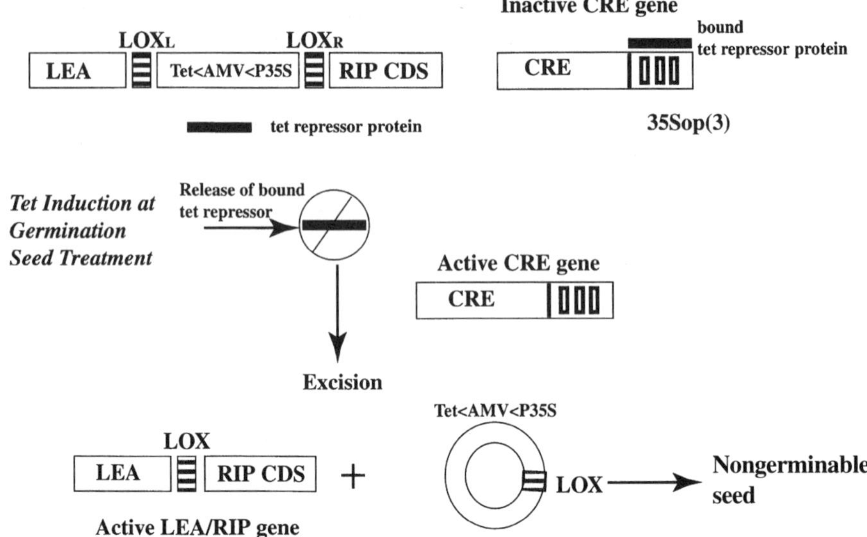

Fig. 1. Schematic cartoon of the Technology Protection System. LEA, Late Embryogenesis Protein gene promoter; LOX, recognition site for the CRE recombinase; Tet, tetracycline; CRE, CRE site-specific recombinase coding sequence; RIP CDS, Ribosomal Inhibitory Protein coding sequence.

of the TPS system have been assayed for functionality. We have demonstrated that plants derived from tetracycline-treated seeds exhibit no visible effect from either the switching on of the TPS system or treatment of the seed with tetracycline. All look normal and all produce normal yields of seeds. Polymerase chain reaction (PCR)-based analysis of DNA fragments generated from these plants reveals that excision occurs, and the LEA promoter–driven germination disruption gene is faithfully generated. Northern analyses of seedlings derived from seeds treated with varying levels of tetracycline indicate that, at between 50 and 100 μg/mL tetracycline, excision of the tet repressor stuffer region occurs in the majority of the cells of the seedlings. That is to say that RNA samples derived from individual whole seedlings do not exhibit any detectable accumulation of tet repressor transcript. These data indicate that the tet repressor/tet promoter system and CRE/LOX recombination is functional and highly effective in tobacco and that the TPS system is functional to this point. We are still in the process of analyzing the germinability of the seeds that result from these plants.

In cotton, we have plants that contain each of the two types of constructs, but not yet in the full combination form of TPS. We are currently starting the process of producing homozygous (same gene on both chromosomes and producing similar traits from generation to generation) lines in order to start the crosses that will generate TPS cotton plants. It is in cotton that we have the best hope of seeing the seed

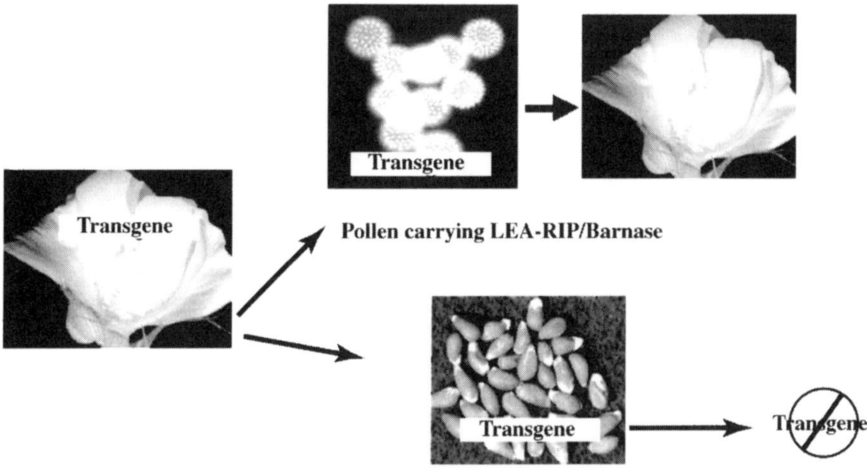

Fig. 2. Schematic of the biosafety aspects of the Technology Protection System.

nongerminability aspect of TPS because the LEA promoters we have used were derived from cotton (it is unclear as yet whether cotton LEA promoters are active in tobacco). Future trials, once the system has been tested in the laboratory, will also focus on environmental and human health safety issues.

Advantages of TPS

The obvious advantage of TPS and one of its least-mentioned benefits addresses a primary biosafety issue. When TPS is activated in a crop, it has the potential to prevent the escape of transgenes, such as genes for herbicide or pest resistance (incorporated into a TPS-containing line), into the surrounding environment through hybridization with closely related weed species. The development of technology designed to prevent gene escape was stated as a major priority for plant biotechnology in a recent report from the National Academy of Sciences National Research Council (http://national-academies.org). Prevention of transgene escape is achieved by two mechanisms, the lack of germination of volunteer seed and the generation of nongerminable seed resulting from fertilization with TPS pollen. This has a supplemental benefit for farmers in that the previous year's crop does not emerge as a contaminating "weed" in a new rotation crop. This is of major importance to farmers who are using herbicide-resistant crops within a rotation program.

TPS will provide a more level playing field for North American farmers because farmers in other countries will also have to pay for varietal and transgenic traits. Under some conditions, TPS may be useful in restricting the distribution of seeds of transgenic varieties. This could be especially useful in preventing planting of transgenic crops in environmentally sensitive areas, or in keeping the seeds out of the hands of unauthorized users. Because of the possibility of a return on investment in breeding research, many more improved varieties, in a broader range of crop species, should become available to producers. This is especially true for crops that have not been given optimum breeding attention (e.g., wheat, rice, or soybeans) or in countries in which breeding research has not been at a level proportionate to the agricultural importance of the crop. In this regard, TPS will also allow for the introduction of transgenic traits into crops in countries in which such technology has not been introduced or used. In this way, all farmers should be the direct beneficiaries of more improved varieties carrying higher yields, improved quality traits, and more and better pest resistance.

TPS will be broadly available to both large and small seed firms. For this reason, it is anticipated that TPS will encourage increased biotechnological and breeding research in many crop species and geographic areas. Genetic diversity in many important crops is a real concern of both private and public breeders today. There is no correlation between TPS and lack of genetic diversity. In fact, with the increased incentive (derived from the use of TPS) for many private seed companies as well as universities to breed crops that have not received sufficient attention in the past, it is entirely possible that diversity will increase as breeders focus on providing unique and improved versions of germ plasm to farmers.

Our future goals are to continue to test TPS in tobacco and to generate and test TPS in cotton. On a planning level, we are looking for ways to improve the system as it stands and to determine the availability of other inducible promoter systems in plants. We have also considered other uses for TPS and how it can be redesigned somewhat to achieve different outcomes, such as for use in open pollination systems, prevention of flowering in forage species, and eradication of noxious weed species.

References

1. Oliver, M.J., N.L.G. Trolinder, D.L. Keim, and J.E. Quisenberry, Control of Plant Gene Expression, U.S. Patent 5,723,765 (1998).
2. Barthelemy, I., D. Martineay, M. Ong, R. Matsunami, N. Ling, L. Benatti, U. Cavallero, M.R. Soria, and D.A. Lappi, The Expression of Saporin, a Plant Ribosome Inactivating Protein from the Plant *Saponaria officinalis,* in *E. coli, J. Biol. Chem. 268:* 6451–6548 (1993).
3. Gatz, C., and P.H. Quail, Tn10-Encoded Tet Repressor Can Regulate an Operator-Containing Plant Promoter, *Proc. Natl. Acad. Sci. USA. 85:* 1394–1397 (1988).
4. Dale, E.C., and D.W. Ow, Intra- and Intermolecular Site-Specific Recombination in Plant Cells Mediated by Bacteriophage P1 Recombinase, *Gene 91:* 79–85 (1990).
5. Odell, J.T., P.G. Ciami, B. Sauer, and S.H. Russell, Site-Directed Recombination in the Genome of Transgenic Tobacco, *Mol. Gen. Genet. 223:* 369–378 (1990).

Chapter 13

Detecting DNA from Genetically Modified Plants

John Fagan

Genetic ID, Incorporated, Fairfield, IA 52556

Introduction

In considering the topic of testing genetically modified organisms (GMOs), the first question that arises is why GMO testing is necessary. The answer is a moving target. At present, the purpose of testing is to differentiate between genetically modified and traditional foods. The need for such differentiation arises from the fact that consumers and governments are demanding that this distinction be made. In the future, however, the need for GMO testing may well be based on altogether different requirements. For instance, many see the future of agricultural biotechnology to be in value-added traits, and consider such testing to be essential for commerce in crops carrying these traits.

Value-added transgenics would, in most cases, be visually indistinguishable from their commodity counterparts. Yet their commercial value would be several times greater, in some cases by orders of magnitude, than that of the corresponding commodity crop. A commodity soybean may be worth \$3/bu, whereas a soybean that produces a specialized, high-value oil or protein composition might be worth \$15/bu. Taking things one step farther, a soybean that produces a costly pharmaceutical could be worth hundreds of dollars per bushel. These value-added crops would be visually indistinguishable from commodities. Thus, it would be necessary to employ highly reliable and sensitive tests that would verify their identity and confirm their commercial value.

The current demand for GMO testing emerges from the dynamics illustrated in Figure 1. The entrance into the marketplace of genetically modified foods, which generates discussion and debate within society triggers this pattern. A consistently observed outcome of this process is increased public awareness of GMOs. This in turn generates consumer demand for choice in the marketplace, which is expressed as a concrete market demand for non-genetically modified products, that is, products that are verifiable as being produced from non-genetically engineered inputs and processes. As this demand stabilizes, the food and agricultural industries perceive this new market demand and respond by providing products that meet the expectations of consumers. Often the final step in this sequence is the establishment of government regulations relating to genetically modified crops.

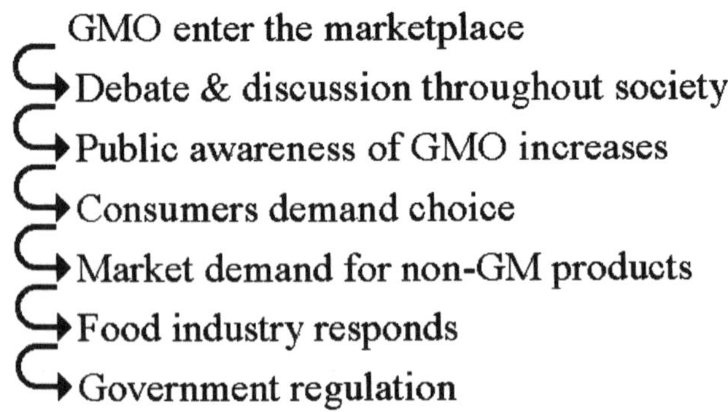

FIG. 1. Emergence of the issue of genetically modified organisms.

The dynamics depicted in Figure 1 constitute a recurring pattern that has unfolded in a consistent manner in country after country around the world. This pattern is in full momentum in major markets for American agriculture. In the European Union, it began in the northern states (UK, Germany, France, Denmark) and is now well entrenched in all of the EU member states, including Italy and Greece. This recurring pattern has also reached a developed stage in Switzerland, Norway, Japan, and Australia. In a number of other regions and countries, this pattern is emerging to one degree or another (Pacific Rim, South Africa, Brazil, and the Middle East). Especially in the last two years, we see this same trend beginning to unfold in the United States and Canada.

Two consistent outcomes of this pattern are market demand for non-genetically modified products and governmental initiatives to regulate genetically modified foods. There are two classes of GMO-related regulations. The first class comprises regulations that govern the use of GMOs. These regulate the cultivation of GMO and their commercialization for human food and animal feed. EU directive (EC)No.90/220 is an example of such a regulation (1) . The second class of regulations governs the labeling of foods produced through gene modification and foods that contain ingredients or other inputs generated using gene modification. EU directives (EC)No.258/97 and (EC)No. 1139/98 are examples of labeling regulations (2,3).

Generally, the importation of genetically modified foods or feed is contingent upon approval of these products for specific uses. National and regional differences in approval status for a given genetically modified crop can create substantial challenges. Table 1 illustrates this situation in the case of corn. The two right-hand columns list all of the transgenic corn events or varieties that have been approved for cultivation in the United States. Of these, the top nine have actually been produced commercially on a large scale. On the right is the approval status of these products for human use in various international markets. Only four of the nine have been approved in the EU and Switzerland, three have been approved in Norway, and eight in Japan. Approval status differences create challenges for

TABLE 1
National Approvals of Genetically Modified Corn

			Approval for human use			
Company	Event	Brand name	EU	Japan	Switzerland	Norway
Novartis/ Myc.	E176	Maximizer	Yes	Yes	Yes	No
Novartis	Bt 11	YieldGard	Yes	Yes	Yes	Yes
Monsanto	Mon810	YieldGard	Yes	Yes	Yes	Yes
AgrEvo	T25	Liberty Link	Yes	Yes	Yes	Yes
AgrEvo	T14	Liberty Link	No	Yes	No	No
DeKalb	DLL25	GR	No	Yes	No	No
DeKalb	DLL418	Bt Xtra	No	Yes	No	No
Monsanto	GA 21	Roundup Ready	No	Yes	No	No
AgrEvo	CBH 351	StarLink	No	No	No	No
Monsanto	Mon 801	Not commercialized	No	No	No	No
Monsanto	Mon 802	Not commercialized	No	No	No	No
Monsanto	Mon 805	Not commercialized	No	No	No	No
Monsanto	Mon 809	Not commercialized	No	No	No	No
Monsanto	839/831/832	Not commercialized	No	No	No	No
Pioneer	676/678/680	Not commercialized	No	No	No	No
Plant Gen. System	MS3	Not commercialized	No	No	No	No

exporters attempting to move corn or corn products into international markets. Not only is it necessary to determine whether genetically modified material is present in order to comply with labeling regulations in these countries, but it is also necessary to ascertain whether a given lot of product contains varieties or events that have not been approved for the specific country of import.

The current status of labeling regulations in the European Union (2,3) exemplifies the situation encountered in many parts of the globe. Products of unknown composition or products that contain >1% genetically modified material must be labeled with a phrase such as "Produced through gene modification." Products containing <1% genetically modified material do not require any label, as long as the producer can provide strong traceability documentation demonstrating that positive efforts were taken to avoid GMO admixture. Testing alone is not enough. The producer must have in place a strong and well-documented identity preservation (IP) system.

To comply successfully with government regulations and respond to market demands, food producers are now using five basic tools. The first is testing to detect the presence of GMO and to quantify the level of genetically modified material, if any is present. IP and traceability systems are a second tool. Quality control/assurance systems that support IP are a third tool. The fourth is third-party certification, which is a system through which the effectiveness of IP and quality control systems can be independently evaluated and verified. Certification by an independent, impartial third party enhances confidence in producer claims. Finally, accurate information on regulatory issues, market trends, and technology is the fifth and essential tool. Of these tools, this paper focuses primarily on testing. However, we will first briefly discuss third-party certification

Certification

The objective of certification is to validate the producer's claims regarding the genetic status of the product. Producers are motivated to take part in certification programs primarily on the basis of their commitments to quality and to fulfilling the customer's needs. Certification is also considered to increase confidence in the producer, inspire customer good will, and differentiate the product in the marketplace. Certification systems are also used as a mechanism with which to ensure and document compliance with GMO-related regulations. This reduces and distributes liability.

Certification often reduces the cost of verifying product quality. For example, if in the absence of certification, several manufacturers were using the same source of non-genetically modified lecithin, under current EU regulations, each manufacturer would be required to audit the production chain for that lecithin all the way back to the farm level. This would be necessary in order for each manufacturer to comply with the EU requirement that labeling claims be based on "all reasonable precautions and all due diligence." In this case, each manufacturer would carry out a separate audit of the supply chain. In contrast, third-party certification of such a supply chain would entail a single, thorough audit whose results could be used by anyone who purchases product from that lecithin producer. Instead of several expensive audits, the cost of a single audit would be distributed among several users. This would reduce costs for both the buyer and the seller, while maintaining, or actually increasing, confidence in the quality of the product.

Yet another ramification of third-party certification is that it facilitates sourcing, whereby the directory of certified producers may be used by any organization downstream in the food chain to discover sources of non-genetically modified inputs required for manufacturing their products. For instance, a manufacturer of non-genetically modified baked goods could scan a directory of certified suppliers to find sources for flour that he or she can be confident are not genetically modified.

GMO Testing

Analytical procedures that are designed to detect the presence of genetically modified materials in food and agricultural products are an important tool for allowing individuals at all levels of the food chain to deal effectively with the GMO issue. An effective GMO analytical system provides data that enable both buyer and seller to enter confidently into a contract that specifies GMO content of the commodity. The fundamental characteristics of such an analytical system are as follows:

- It must detect all GMO present in the marketplace.
- It must provide quantitative information about the level of genetically modified material.
- The system must deliver maximal reliability and reproducibility.
- Such a system must be sufficiently flexible and comprehensive to deliver informative results for a wide variety of foods and agricultural products.

Tests for GMOs are based on methods that detect those biomolecules that are present in genetically modified organisms due to the process of gene modification and that are not present in their conventional counterparts. Gene modification (also called recombinant DNA methods or gene splicing techniques) introduces new genetic information, i.e., new DNA sequences, into the genome of an organism. This new genetic material can be derived from virtually any species, from viruses to microorganisms, to plants or animals. It could also be synthesized chemically. Often the genetic material is stitched together from DNA segments derived from several of these sources. Once introduced into the genome, the transgenic DNA reprograms the cells of the recipient organism to produce new mRNA and proteins. These proteins confer new characteristics or functions upon the organism. In some cases, these new functions involve the synthesis of other biomolecules not normally present in the organism, such as new starches or oils.

In principle, GMO detection methods could measure any of the four classes of gene modification-specific biomolecules, i.e., DNA, mRNA, transgenic proteins, and novel biosynthetic products. In practice, GMO testing methods have not focused on mRNA detection because of the relative instability of these molecules, and, to date, biosynthetic products have not been the targets of GMO tests. Instead, analytical methods have focused on DNA and protein detection. The following sections describe these methods.

Immunological Analysis of Transgenic Proteins

Immunological tests for GMOs have exploited both the enzyme-linked immunosorbent assay (ELISA) and the lateral flow test format (4–6). Although there are many different configurations for ELISA tests, the basic design is as follows: First, immobilized antibodies capture a specific antigen. This antigen-antibody reaction is detected using a second antibody that also reacts with the antigen, and is also linked to an enzyme that catalyzes the conversion of a compound present in the reaction vessel into a second compound that can be quantified colorimetrically. The lateral flow test makes use of the same basic immunochemistry but is configured to allow convenient field analysis with visual assessment of results.

Because they require minimal processing of the sample, immunotests can be completed quite rapidly (4). Moreover, in the lateral flow format, immunotests are very convenient and easy to carry out and do not require sophisticated equipment. The home pregnancy test is a familiar example of the lateral flow test. This format is particularly useful for field GMO tests, in which the truckloads of soybeans or corn can be screened rapidly and inexpensively at the grain handling facility for single genetically modified traits. The speed, convenience, and economy of immunological tests offer substantial utility. However, the limitations of this method should be recognized to ensure appropriate application. One crucial limitation is in the area of quantification.

Although ELISA tests can be configured to function quantitatively, in the context of GMO testing, this capacity for quantification cannot be used advantageously. This is because it is difficult to translate mass of transgenic protein, measured in the sample extract, into the percentage GMO, which is the quantitative basis for many national regulations on genetically modified foods. For an example, see EU directive 258/97 (2). The percentage GMO refers to the weight percentage of food derived from genetically modified materials. For example, a truckload of 20% genetically modified corn might contain 1 t of transgenic corn and 4 t of conventional corn. If one were to conduct quantitative ELISA on a representative sample of that corn, the analysis would provide information, with reasonably good precision on the mass (nanograms) of a specific transgenic protein, such as Cry1Ab, extracted from a given number of grams of corn. The difficulty comes in accurately extrapolating from this value to the percentage GMO because there is no constant relationship between these two parameters (mass of transgenic protein extracted and mass of corn grains).

If the level of expression of the transgenic protein were constant, i.e., if the nanograms of transgenic protein expressed per gram of transgenic corn were constant, then one could compare the result of this analysis to a series of standards containing known amounts of transgenic corn. The percentage GMO could in this case be estimated on the basis of that comparison. However, expression is not constant. It is influenced by weather, soil, and other cultivation conditions.

For example, Roundup Ready soybeans express transgenic EPSPS at levels ranging from 0.179 to 0.395 ng/mg. This is more than a twofold difference. In virtually every instance, the standards used for calibrating the analysis will be derived from a different lot of soybeans than the sample. Therefore the level of expression in the sample will differ from the standards, and it will not be valid to estimate the GMO content of the sample by comparison with those standards. A sample judged to contain 1% genetically modified soybeans on the basis of such a comparison could contain as little as 0.5% or as much as 2%.

In addition, different transgenic events express the same transgenic proteins at widely varying levels. For example, E176, Bt11, and Mon810 all express transgenic Cry1A proteins, but they do so at widely different levels. If the truckload of corn contained a mixture of two or three of these, immunological analysis could be used to determine the amount of transgenic Cry1A protein present per gram of corn, but it would be impossible to extrapolate from this measurement to the percentage GMO present in the truckload. This is because the analyst will not know whether the sample is comprised of a single variety or a mixture, nor will he know the relative proportions of the varieties or events that may be present. Another factor that influences quantification is efficiency of extraction. If the sample and standards are not ground to the same mesh size and extracted for the same length of time, the transgenic proteins will be extracted with different efficiencies in standards and sample.

In summary, as a result of several confounding factors, the amount of a transgenic protein present in, or extractable from, a grain of food is variable and cannot be used as a measure of the proportion of that food that is transgenic.

A second limitation of immunoassays is that the transgenic proteins expressed in some genetically modified crops are not detectable by immunoanalysis. For example, the glyphosate-resistant corn variety GA21 expresses a transgenic EPSPS protein that differs from the native corn EPSPS by only two or three amino acids (7). The structures of the transgenic and native EPSPS proteins are so similar that all attempts to develop antibodies capable of differentiating the two have been unsuccessful. Thus, to date, no immunotest exists that is capable of detecting this transgenic event.

A third limitation of immunoassays is that transgenic proteins are easily denatured during food processing, either destroying the ability to recognize these proteins with immunological reagents or reducing sensitivity to detection (5,8). Thus, detectability is variable and process dependent, again compromising the utility of immunological quantification methods. As stated recently (5), matrix-matched standards would be required for quantification. Not only would it be necessary to process the standard material under conditions identical to those of the sample, but the proportions of different genetically modified events comprising the standard would have to necessarily match those of the sample. These are conditions that can be fulfilled economically in only a small fraction of the circumstances in which it is necessary to quantify GMO content.

Despite limitations, immunological tests serve a useful role. Their application at early stages of the chain is well accepted at this time, especially the points at which rapid field tests are needed. For instance, they are often used in checking trucks before they unload their cargoes at grain-handling facilities. The initial triage that these tests provide prevents the introduction of truckloads of corn or soybeans that contain high levels of genetically modified material into silos designated for non-genetically modified product. ELISA is also being used for quantification in situations in which economy and convenience are considered more critical than accuracy.

Analysis of Transgenic DNA by PCR

The polymerase chain reaction (PCR) can be used in genetics-based analysis of GMOs (8–21). PCR uses biochemical processes to scan through a sample of DNA and to locate one or more specific DNA sequences, called target sequences. The target sequence is then amplified billions of times, making it possible to detect that target sequence with high sensitivity and also to quantify the proportion of DNA molecules in the sample that contain that target.

The key elements in the PCR process are the following: (i) "primers," which are small DNA molecules whose sequences correspond to the specific DNA sequence, called the target sequence, for which one wishes to search; (ii) a heat stable enzyme, typically Taq polymerase, which synthesizes new copies of the target sequence when primers interact with that target sequence; and (iii) a thermocycler, which is an apparatus that can be programmed to carry the contents of the PCR reaction vessel through multiple, precisely controlled temperature cycles.

During the PCR process, the primers locate and bind to a specific target sequence, if it is present within the DNA sample. In the presence of Taq polymerase and other essential reaction components, the thermocycler carries the reaction through 20–40 temperature cycles, each of which doubles the number of copies of the target sequence. In 40 PCR cycles, this exponential amplification generates 10^{12} DNA molecules for every target sequence originally present. Each of these is a copy of the target sequence originally detected by the primers.

Because of the powerful amplification that occurs during PCR, this process is a highly sensitive detection method. Because the interactions between the primer and target DNA molecules are highly selective, the PCR amplification process is highly specific. Another advantage is that PCR is capable of detecting all GMOs because, even if the transgenic protein is not expressed in the food part of the plant or even if the transgenic protein is indistinguishable from the native protein by immunoanalysis, the transgenic DNA will be present and can be detected by PCR. DNA is less subject to denaturation and degradation during food processing than are most transgenic proteins. Thus, even when transgenic proteins have been degraded to the point where immunotests are ineffective, PCR analysis can, in most cases, still successfully detect the presence of genetically modified material (8,10). The robust nature of this method makes it possible to use PCR to test for the presence of genetically modified material at almost all points in the food chain, from the farmer's field to the consumer's plate. PCR can also be used to quantify GMO content in almost all food products, including highly processed food. The only exceptions are the most highly modified food ingredients, such as certain chemically modified starches, the most highly refined grades of vegetable oil, and highly fermented products such as soy sauce.

One of the most significant advantages of PCR-based GMO analysis lies in the area of quantification (8,15). Because the DNA extracted from a sample contains not only the transgene, but also all of the other naturally present genes of the organism, and because the copy number of most genes, including the transgene, are invariant, it is possible to design PCR analytical procedures that provide definitive determinations of the percentage of genetically modified material present in the sample. In the simplest form of this strategy, DNA is isolated from a statistically representative subsample of the sample and is subjected to PCR analysis using two primer sets, one designed to detect a transgenic sequence (called the transgene) and one designed to detect a gene that is present in all varieties of the crop of interest (called the species-specific reference gene).

The percentage GMO can be calculated from quantitative PCR measurements of the transgene and the species-specific reference gene (SSRG). Corn is used as an example here. Because the SSRG is present in the same copy number in both transgenic and conventional corn, the PCR signal derived from these sequences can be used as a measure of the total amount of corn present in the sample. Similarly, the PCR signal derived from the transgene can be used as a measure of the total amount of transgenic corn present in the sample. Thus, the ratio of these two signals is a good measure of the percentage of a given sample that is transgenic; it is an accurate measure of the percentage GMO of the sample. Empirical assessments indicate that this genetic ratio is generally within 15% of the actual mass ratio of transgenic to total (transgenic plus conventional) corn. This variation is due to influences such as variation in water and oil content of the grain.

Thus, by quantifying the SSRG in parallel with the transgene, we obtain an internal and self-consistent reference point that makes it possible to relate the measured level of the transgene to the total amount of corn in the product. This analytical approach, linked with the ability to devise primer sets that differentiate definitively among all transgenic events or varieties (see below), makes it possible to quantify GMO content, i.e., percentage GMO, in a precise manner that is based on first principles of the genetics of the sample.

Because the transgene and the SSRG are present in the same pool of genomic DNA within each cell of the corn kernel, these two genes will be subjected to the same degradative and other effects during food processing and during DNA isolation, and they will be extracted with equal efficiency during DNA isolation. Because of this, the SSRG can serve as an internal control that allows the analyst to correct for variations in extraction efficiency and possible degradation effects.

PCR also has significant limitations. The most prominent include the following: (i) analysis time is lengthy, compared with immunotest methods; (ii) PCR is complex and technically challenging; and (iii) the high sensitivity of PCR can present problems when this method is placed in the hands of inadequately trained technicians. The technical complexity of this method requires substantial overhead in highly trained personnel and expensive equipment. For instance, the most commonly used instrument in real-time PCR retails for >US$100,000. Because of these cost constraints, PCR is applicable only to material lots large enough or valuable enough to justify significant testing costs.

The lengthy time required for PCR analysis limits its applicability and makes it necessary to integrate this method into the production, storage, and transport chain at points at which time constraints are not problematic. For instance, it is simply not practical to use PCR to screen trucks of soybeans or corn as they are unloaded at a grain handling facility. Lateral flow immunotests are preferable in this case. On the other hand, once grain is placed in extended storage on farms or at the local grain-handling facility, the sensitivity of PCR can be used advantageously, facilitating the compositing of samples from several silos for a single analysis. This strategy can reduce costs under most conditions in which a testing turnaround time of a few days will not interfere with the operation. The higher sensitivity of PCR can also present a technical challenge. Its sensitivity makes it necessary for analytical personnel to undergo extensive training, and for stringent controls, standards, and precautions to be incorporated into the analytical procedure.

In considering GMO quantification, two other limitations of PCR have been discussed. First, ubiquitous transgenic sequences are commonly used to screen for GMOs because primer sets for only a few such sequences are sufficient to detect all genetically modified crop varieties. However, use of these ubiquitous transgenic sequences for quantification is problematic because they are present in different copy numbers in different genetically modified events or crop varieties. A sample of unknown origin could contain any single event, or could contain several. In such a situation, the variable copy number of these ubiquitous sequences makes it impossible to use them to quantify GMO content definitively.

For instance, the 35S promoter from the cauliflower mosaic virus is used in all except one transgenic soybean and corn varieties that have been approved for commercialization in the United States. Thus, primer sets that recognize this sequence element are useful for screening. However, this sequence element is not an ideal PCR target for quantitative PCR of corn, because its copy number varies widely from event to event, from 0 to 4 or more copies.

Analysts will not normally know which transgenic events are present in any particular sample. Therefore, they cannot take into account the differences in copy number for a target sequence such as the 35S promoter. They will simply obtain an overall value for the target sequence that they can quantify against a standard for one event. For example, a sample might contain 0.5% CBH351, 0.2% Bt11, and 0.2% GA21. Analysis based on the 35S promoter, using Bt11 as standard, would indicate that the sample contained 1.2% genetically modified material. The 0.5% of CBH351 would register at 1.0% because in CBH351, the copy number for 35S is 4. The 0.2% of Bt11 would be quantified correctly, and the GA21 would not be detected. Similar problems arise using the No sterminator or other common sequence elements as the basis for quantification.

This is a problem with GMO quantification as it is commonly carried out because sequence elements such as the 35S promoter are generally used. However, this problem has been solved by several laboratories that have based quantification not on primer sets that recognize ubiquitous sequences such as the 35S promoter, but on primer sets that recognize sequences specific for single transgenic events, and that are, therefore, specific for those events. In the example above, instead of quantifying on the basis of a primer set specific for the 35S promoter, we would use several primer sets, one specific for each corn event that has been commercialized. These would first reveal that CBH351, BT11, and GA21 were present and, second, provide separate quantifications for each of these events. Clearly, this requires more effort than quantification via the 35S promoter, but results are definitive.

The second limitation of PCR-based quantification that has been discussed relates to the complex zygosity of certain crops. This issue is not relevant to soybeans or other current transgenic crops, except for corn. However, other examples could arise in the future. The embryo of a corn kernel is diploid, with equal genetic contributions from paternal and maternal parents, whereas the endosperm of the kernel is triploid, with two identical genomic copies coming from the mother and one from the father. On the surface, this might suggest that the measured GMO content of corn derivatives such as starch, which are derived from the endosperm, would be different from the measured GMO content of the kernel as a whole. Closer consideration, however, reveals that this would be the case for *hybrid seed* generated from homozygous parents because every kernel would have identical genetics, but would not be the case for *grain* derived from such seed.

As a result of the statistics of trait segregation, the contribution of maternal and paternal genetics to the genetics of the grain population is 1:1, both for the embryo and for the endosperm. Any given kernel of that grain population will have a ratio of maternal to paternal genetics of 2:1, but within the grain kernel population as a whole, the ratio will be 1:1 for any given trait. Therefore, al! derivatives, whether produced from embryo or endosperm, will have the same average genetics, and the percentage GMO determined for any derivative will be consistent with the value obtained for unprocessed corn.

A point of caution does emerge from this analysis, however, i.e., the genetics of the *hybrid seed* differ from those of the *grain* derived from that seed. The former are biased toward maternal genetics, due to the higher maternal gene dosage in the endosperm, whereas in the grain derived from such a hybrid, no bias exists within the average genetics of the kernel population. This indicates that one must select appropriate standards for a given analysis, using grain standards when analyzing corn grain, and hybrid seed standards when analyzing hybrid seed. Alternatively, a seed standard (or *vice versa*, a grain standard) can be characterized empirically as to copy number, and a correction factor applied when it is used in quantification of grain samples (or *vice versa*, seed samples).

In summary, both immunomethods and PCR methods can be used effectively to screen qualitatively for GMOs, and each of these approaches offers advantages under certain circumstances. In the area of quantification, however, PCR provides definitive quantification, whereas immunomethods do not.

PCR-Based GMO Analysis-Steps of the Process

PCR analysis of GMOs involves the following five steps: sample preparation, DNA purification, target amplification, detection of reaction products, and interpretation of results. These are discussed in the following paragraphs.

Sample Preparation. Sample preparation is a key element in PCR testing. Care must be taken to ensure that the sample taken is representative of the lot of product and that the analytical sample derived from the larger field sample is also representative. The field sample must be large enough to be representative of the lot of material of interest, and it must be taken in a manner that ensures representation from all parts of the lot. The field sample must also contain a sufficient number of units (in the case of grain, kernels) to ensure that the analysis will be statistically robust at the limits of detection and quantification relevant to the assay. For PCR, the limit of detection is typically 0.01% and the limit of quantification is typically 0.1%. To exploit this sensitivity fully, the field sample should contain at least 10,000 units. Similarly, the field sample should be processed and the analytical sample taken in a manner that ensures that the analytical sample is representative. For instance, reducing a 10,000 bean sample of soy to fine flour and taking two 1-g analytical samples from the homogenized sample will ensure full quantitative representation of the original 10,000 beans in the analytical samples. This is because thousands of particles from each of the original beans will be present in the final analytical sample.

DNA Purification. Purification procedures that produce DNA free from PCR inhibitors, as well as minimize DNA degradation and achieve good yields, are key to gaining reliable and informative results. Food products vary tremendously in their physical and chemical compositions. For this reason, it is essential to customize DNA extraction methods to function optimally for each food matrix. Clearly, DNA extraction kits purchased from a scientific supply house are unlikely to perform adequately for all sample types.

PCR Amplification. The PCR amplification process was described above. The ability of PCR to amplify a target sequence billions of times is the basis of the utility of this method. Normally, between 30 and 40 cycles of PCR amplification are carried out, yielding signals that are easily observed even if the original target sequence is present in only one or a few copies. With this method, it is possible to routinely detect the presence of as little as 0.01% genetically modified material.

Detection of PCR Reaction Products and Interpretation of Results. Products of the PCR amplification process can be detected by several different methods. The two most common are electrophoretic analysis, in which the amplified molecules appear as distinct bands on an agarose or acrylamide gel, and fluorometric analysis, in which the PCR process is modified to generate fluorescent products in proportion to the number of amplification events that take place. A diagrammatic example of electrophoretic analysis is found in Figure 2A. The top panel presents the results from analysis of two samples. Ten separate PCR reactions were run for each sample, two with each of five distinct primer sets.

The first primer set is an internal control that indicates whether PCR inhibitors may be present. If inhibitors are not present, the intensity of the electrophoresis bands from the control will match the intensity of the internal standard control that

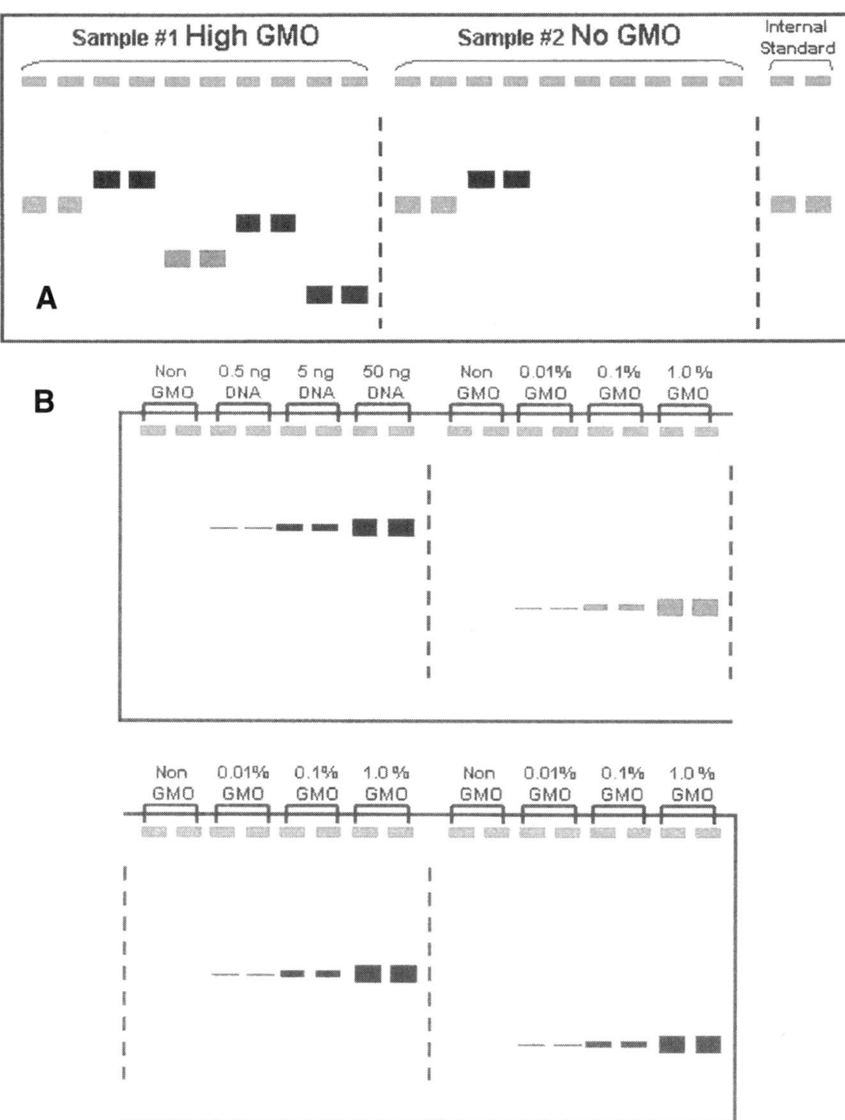

FIG. 2. Triple-check semiquantitative polymerase chain reaction (PCR) analysis: (A) two samples; (B) external standards series.

is at the far right portion of the panel. The second primer set is designed to determine whether the DNA sample is intact and free from degradation. This primer set is specific for a reference gene for the species of interest, i.e., the genetically modified primary set. For example, to test for genetically modified soybeans, a primer set would be selected that targets a gene known to be present in all soybeans, whether transgenic or conventional. Many laboratories use a primer set specific for the soy lectin gene for this purpose (10,15). The reaction products generated using this primer set are compared with those in the first set of reactions found to the left-hand side of Figure 2B. If the DNA sample is free from degradation, then the intensity of the bands will correspond to the highest intensity bands in that series of standards.

The final three primer sets are specific for three different transgenic sequences. The intensities of the signals generated for a given sample using these three primer sets (Fig. 2A) can be compared with the respective series of external standards (Fig. 2B) to obtain a semiquantitative assessment of the GMO content of the sample. To provide quantitative information, all duplicates must agree, the two control primer sets must show absence of inhibitors and DNA degradation, and all three GMO-specific primer sets must yield consistent results. This example illustrates that more than 30 data points must be considered in determining the GMO content of any given sample, i.e., a total of 10 PCR reactions are carried out with DNA derived from the sample, and more than 20 additional reactions are conducted with external reference standards.

Quality Assurance Measures. Incorporated into the procedure illustrated in this example are five elements designed to ensure the accuracy and consistency of results, including the following: (1) A positive control primer set that recognizes a species-specific reference gene is used to assess whether the DNA is intact and free from degradation. (2) An internal standard primer set and template are incorporated into the assay to test for the presence of PCR inhibitors in the DNA preparation. (3) Multiple GMO-specific primer sets are used to demonstrate reproducibility. (4) Each sample is analyzed in duplicate from start to finish. These duplicates do not originate at the PCR stage of the analysis. Duplicate analytical samples are taken from the homogenized field sample and processed in parallel throughout the analytical procedure. (5) An extensive set of quantitative external standards is employed.

In addition, a number of other measures that are not apparent from this example are essential. These include the following: (i) the PCR reaction conditions and the primer sets used must be optimized for sensitivity and specificity; (ii) DNA purification procedures must be customized for each food matrix to ensure freedom from inhibitors and to minimize degradation; (iii) to reduce the risk of cross-contamination, steps of analysis must be physically and operationally isolated; (iv) stringent quality control measures must be implemented for all analytical procedures and reagents.

When these or similar measures are employed, highly accurate and consistent results can be obtained. For instance, the Joint Research Council of the EU conducted a ring trial in 1998 (12), in which the frequency of false positives for analysis of soybeans using primers for the 35S promoter was 2.1%. The frequency of false negatives was 5.1% for samples that contained 0.1% genetically modified material, but no incorrect results were reported for samples containing 0.5 or 2%

genetically modified material. For both false positives and false negatives, incorrect results were reported by only 2 of the 25 laboratories. Thus, the vast majority of laboratories performed perfectly on all analyses.

Real-Time PCR

The existence of thresholds for the GMO content of products, whether mandated by government regulation or contractual agreement between buyer and seller, necessitates methods for quantifying GMO content of foods and agricultural products. Consequently, earlier qualitative GMO detection methods have given way to methods that offer increasingly more robust quantification. Conventional PCR methods, such as that described above, that are calibrated using external standards or internal competitive templates are, and will continue to be used to provide economical semiquantitative information (8,9,22,23). However, real-time quantitative PCR methods are increasingly the method of choice (8,15).

Conventional PCR measures the products of the PCR reaction at a single point in the reaction profile, whereas real-time PCR methods generate fluorescent reaction products that can be monitored continuously to follow the time course of several PCR reactions simultaneously (24,25). With this method, the complete reaction profiles for as many as 96 samples can be followed in parallel (26). By comparing the profiles of the sample with those of a series of standards, it is possible to determine the GMO content of the sample. Conventional PCR can be used semi-quantitatively over only a limited range of DNA concentrations, whereas real-time PCR provides good quantification over four to five orders of magnitude (24–26). For samples containing at least 0.5% genetically modified material, precision is ± 15–20%. At lower levels, precision is less, but useful quantifications can be achieved down to 0.1% of genetically modified material.

Summary

In many nations around the world, both consumers and government are demanding greater certainty in distinguishing genetically modified and traditional foods. In the future, the demand may also emerge for testing to differentiate among value-added transgenics from conventional products. Today, there are several tools available to respond effectively to the global demand for non-genetically modified food. There are identity preservation (IP) and traceability systems that can operate at every level from seed production, through shipping and storage, to ingredient and finished food manufacture, to retail grocery chains. Strong quality control and quality assurance systems ensure and verify the successful operation of IP systems. Third-party independent certification is increasingly applied as a technology to verify objectively that the product of interest meets a specified production or content standard. Finally, analytical procedures, which are designed to detect the presence of genetically modified materials in food and agricultural products, are an important tool, allowing organizations at all levels of the food chain to ascertain the genetically modified content of food ingredients. This paper has focused on GMO testing methods, outlining the primary strengths and limitations of both immunology- and genetics-based tests, pointing out the areas in which both of these methods can and are being used effectively in the marketplace today.

Acknowledgment

The expert editorial assistance of Randy Coplin is gratefully acknowledged.

References

1. EC Council Directive of 23 April 1990.
2. EC Council Regulation (EC)No 258/97 of 27 January 1997 Concerning Novel Foods and Novel Food Ingredients.
3. EC Council Regulation, (EC)No. 1139/98 of 26 May 1998.
4. Lipton, C.R., J.X. Dautlick, C.D. Grothaus, P.L. Hunst, K.M. Magin, C.A. Mihaliak, F.M. Rubio, and J.W. Stave, Guidelines for the Validation and Use of Immunoassays for Determination of Introduced Proteins in Biotechnology Enhanced Crops and Derived Food Ingredients, *Food Agric. Immunol. 12:* 153–164 (2000).
5. Lipp, M., E.Anklam, and J.W. Stave, Validation of an Immunoassay for Detection and Quantitation of a Genetically Modified Soybean in Food and Food Fractions Using Reference Materials: Interlaboratory Study, *J. Assoc. Off. Anal. Chem. Int. 83:* 99–927 (2000).
6. Stave, J., Detection of New or Modified Proteins in Novel Foods Derived from GMO—Future Needs, *Food Control 10:* 367–374 (1999).
7. U.S. Food and Drug Administration, Safety and Nutritional Aspects of BtCry9C Insect Resistant, Glufosinate Tolerant Maize Transformation Event CBH-351, Food and Feed Assessment Summary submitted by AgrEvo USA, March 3, 1998.
8. Hubner, P., E. Studer, and J. Luthy, Quantitation of Genetically Modified Organisms in Food, *Nat. Biotechnol. 17:* 1137–1138 (1999).
9. Hubner, P., E. Studer, and J. Luthy, Quantitative Competitive PCR for the Detection of Genetically Modified Organisms in Food, *Food Control 10:* 353–358 (1999).
10. Jankiewicz, A., H. Broll, and J. Zagon, The Official Method for the Detection of Genetically Modified Soybeans (German Food Act LMBG 35), *Eur. Food Res. Technol. 209:* 77–82 (1999).
11. Lebensmittelverordnung (Swiss Food Ordinance) of 1 March, 1995, SR 817.02. *Eidgenossische Drucksachen und Materialzentrale,* CH-3003 Bern, 1995.
12. Lipp, M., P. Brodmann, K. Pietsch, J. Pauwels, and E. Anklam, IUPAC Collaborative Trial Study of a Method to Detect Genetically Modified Soy Beans and Maize in Dried Powder, *J. Assoc. Off. Anal. Chem. Int. 82:* 923–928 (1999).
13. Meyer, R., Development and Application of DNA Analytical Methods for the Detection of GMOs in Food, *Food Control 10:* 391–399 (1999).
14. Shirai, N., K. Momma, S. Ozawa, W. Hashimoto, M. Kito, S. Utsumi, and K. Murata, Safety Assessment of Genetically Engineered Food: Detection and Monitoring of Glyphosate-Tolerant Soybeans, *Biosci. Biotechnol. Biochem. 62:* 1461–1464 (1998).

15. Vaitilingom, M., H. Pijnenburg, F. Gendre, and P. Brignon, Real-Time Quantitative PCR Detection of Genetically Modified Maximizer Maize and Roundup Ready Soybean in Some Representative Foods, *J. Agric. Food. Chem. 47:* 5261–5266 (1999).
16. Vollenhofer, S., K. Burg, J. Schmidt, and H. Kroath, Genetically Modified Organisms in Food-Screening and Specific Detection by Polymerase Chain Reaction, *J. Agric. Food Chem. 47:* 5038–5043 (1999).
17. Wassenegger, M., Application of PCR to Transgenic Plants, *Methods Mol. Biol. 92:* 153–164 (1998).
18. Greiner, R., U. Konietzny, and K.D. Jany, Is There Any Possibility of Detecting the Use of Genetic Engineering in Processed Foods? *Z. Ernaehrwiss. 36:* 155–160 (1997).
19. Hemmer, W., Food Derived from Genetically Modified Organisms and Detection Methods, Agency for Biosafety Research and Assessment of Technological Impacts of the Swiss Priority Program Biotechnology of the Swiss Science Foundation, Clarastrosse 13 CH-9058 Basel, Switzerland. (BATS Report 2) (1997).
20. Hemmer, W. and U. Pauli, Labeling of Food Products Derived from Genetically Engineered Crops, *Eur. Food Law Rev. 1:* 27–38 (1998).
21. Hoef, A.M., E.J. Kok, E. Bouw, H.A. Kuiper, and J. Keijer, Development and Application of a Selective Detection Method for Genetically Modified Soy and Soy-Derived Products, *Food Addit. Contam. 15:* 767–774 (1998).
22. Studer, E., C. Rhyner, J. Luthy, and P. Hubner, Quantitative Competitive PCR for the Detection of Genetically Modified Soybean and Maize, *Z. Lebensm.-Unters.-Forsch. (Food Res. Technol.) 207:* 207–213 (1998).
23. Hardegger, M., *Eur. Food Res. Technol. 209:* 83–87 (1999).
24. Desjardin, L. E., Y. Chen, M.D. Perkins, L. Teixeira, M.D. Cave, and K.D. Eisenach, Comparison of the ABI 7700 System (TaqMan) and Competitive PCR for Quantitation of IS6110 DNA in Sputum During Treatment of Tuberculosis, *J. Clin. Microbiol. 36:* 1964–1968 (1998).
25. Heid, C.A., J. Stevens, K.J. Livak, and P.M. Williams, Real Time Quantitative PCR, *Genome Res. 6:* 986–994 (1996).
26. Livak, K.J., Quantitation of DNA/RNA Using Real-time PCR Detection, Perkin-Elmer's website (http://www2.perkin-elmer.com), 1996.

Index